Beauty from the Beast

Beauty from the Beast

Plate Tectonics and the Landscapes of the Pacific Northwest

Robert J. Lillie, PhD
Certified Interpretive Trainer
Emeritus Professor of Geosciences
Oregon State University

Wells Creek Publishers
Philomath, Oregon

← Cape Perpetua Scenic Area, Oregon. Lava flows formed in the Pacific Ocean are now part of the North American continent. (Photo by Robert J. Lillie).

Designed by Robert J. Lillie.

Photos and illustrations (unless otherwise noted) by Robert J. Lillie.

ISBN 978-1-512-21189-4

Published by Wells Creek Publishers, Philomath, Oregon.

Printed by CreateSpace.

Available from Amazon.com and other retail outlets.

Manufactured in the United States of America.

Published 2015.

Cover: The beast of an exploded volcano, Mt. Mazama, is now the beauty of Crater Lake. (Mt. St. Helens background photo from U. S. Geological Survey. Crater Lake photo by Robert J. Lillie).

To Barb, my kindred spirit and best friend.

Acknowledgments

Beauty from the Beast is the culmination of my work as a geologist, college professor, park ranger, and interpretive trainer. Over the decades I have been inspired and encouraged by many organizations and individuals to explore the natural world and relate it to both scientists and non-scientists. I hope that the photos, illustrations and demonstrations in this book will convey the "Beauty from the Beast" story of the Pacific Northwest in ways that are meaningful to park visitors, educators, and others inspired by the region's mountains, valleys, and coastlines.

I wish to thank Oregon State University for giving me the opportunity to explore mountain ranges around the world and to teach students about the wonders of geology. The National Park Service (NPS) provided funding and opportunities for my students and me to work in parks and develop geology training materials for park rangers. I am especially grateful to Judy Geniac, Jim F. Wood, and Jason Kenworthy from the NPS Geologic Resources Division for their unselfish help and encouragement over the years.

The EarthScope Program of the National Science Foundation provided funding for me to collaborate with park and museum interpreters around the country, and to work with Northwest educators on earthquake and tsunami science and preparedness. I am also grateful to the U. S. Forest Service, the Bureau of Land Management, and Oregon State Parks for opportunities to work with their staffs in geology interpretive workshops. Jason O'Brien made it possible for me to develop workshops and training materials for the Oregon Master Naturalist program. And I'm grateful to the National Association for Interpretation for helping me find ways to relate geology and its deeper meanings to the public.

The original manuscript and illustrations benefited greatly from thoughtful reviews by Daniele McKay, Ray Pilgrim, Jason O'Brien, and Stewart Holmes. My good friend Steve Mark, the historian at Crater Lake National Park, provided much inspiration through his insightful conversations over the years, aided by his exceptional home brews.

My wife Barb has been especially helpful with her encouragement and artful eye during discussions about the book cover, writing and illustrations.

Contents

Introduction

Plate Tectonics:
Beauty from the Beast

The inspiring landscapes of the Pacific Northwest captivate with their beauty and power—and sometimes danger! But the same geological forces that threaten our lives with earthquakes and volcanic eruptions also nourish our spirits by forming the region's spectacular mountains, valleys, and coastlines. National, state, and local parklands highlight the scenery and reveal Earth's processes in action.

Northwest landscapes display a variety of geological processes that have had profound impacts on climate, culture, and commerce. Native American oral tradition and modern geological studies reveal that volcanic eruptions, earthquakes, and tsunamis have at times devastated the region, and that more are sure to come. So why do people continue to live in and visit the Pacific Northwest?

Northwest residents and visitors have learned to live with these geological beasts, in part because the beauty is so inspiring. There is no place on Earth that is immune to natural disasters. Whether it's hurricanes on the Gulf

Beauty from the Beast: Volcanoes

Mt. St. Helens
National Volcanic Monument, Washington

U. S. Geological Survey

Crater Lake
National Park,
Oregon

Robert J. Lillie

The 1980 eruption of Mt. St. Helens unleashed the beast of death and destruction. But volcanic forces are also responsible for the breathtaking beauty of Crater Lake.

← Silver Falls State Park, Oregon. Lava flowed more than 300 miles from its source in the Columbia Plateau to the Willamette Valley region. (Photo by Robert J. Lillie).

Beauty from the Beast: Earthquakes

Copalis River, Washington

U. S. Geological Survey

Olympic National Park, Washington

Robert J. Lillie

Trees along the Washington coast were killed by salt-water invasion when the land suddenly dropped during the last great Cascadia Subduction Zone earthquake in the year 1700. The same forces that shift the land so dramatically also gradually build spectacular landscapes like the coastline of the Olympic peninsula.

Beauty from the Beast: Tsunamis

Beverly Beach State Park, Oregon

Robert J. Lillie

Four children were killed at Beverly Beach in Oregon when they were overcome by giant waves—a tsunami—generated by the 1964 Great Alaska Earthquake. The serenity of the same beach attracts tens of thousands of visitors to Beverly Beach State Park every year. Our past experiences have helped us develop emergency measures, as depicted in this tsunami evacuation sign, so that we can more safely live with the beast while enjoying the beauty.

Coast, tornadoes in the Midwest, or earthquakes in California, communities learn to understand and live with the dangers posed by natural hazards. The key is to develop lifestyles and land-use policies that minimize the risks posed by geological hazards, and to devise disaster preparedness plans that minimize the damage when geological events occur. And equally important, national parks and other special places help us appreciate the beauty that the same geological forces create over time.

Plate Tectonics

The concept of plate tectonics helps us understand how Northwest landscapes formed in the past and how they continue to change during our lifetimes. Plate-tectonic processes result in earthquakes, volcanic eruptions, tsunamis, landslides, and other geological hazards that affect our lives and livelihoods. But the same processes have built magnificent landscapes—Crater Lake, the Columbia Plateau, Mt. Rainier,

the Olympic Mountains, San Francisco Bay, and Yellowstone, to name just a few—that humans and a dynamic Earth continuously modify. Residents and visitors to the Pacific Northwest and surrounding regions are in unique positions to witness plate tectonics in action, and to learn strategies to co-exist with a dynamic planet.

The Earth is layered because it consists of different chemical materials. But that's only part of the story. If you could descend to a great depth within the Earth, you would be incinerated by scorching temperature and crushed by enormous pressure. Those conditions harden and soften materials, and result in an outer shell of tectonic plates that ride over a softer layer below. Much of the action on Earth's surface—earthquakes, volcanic eruptions, and the formation of mountain ranges—occurs along the boundaries of the moving plates.

Volcanic activity occurs at a **divergent plate boundary** where the hot, softened mantle ("asthenosphere") rises and starts to melt. But there are only shallow earthquakes because the

Nature of the Inner Earth

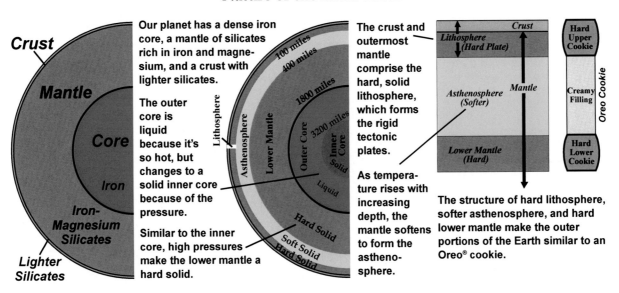

Our planet has a dense iron core, a mantle of silicates rich in iron and magnesium, and a crust with lighter silicates.

The outer core is liquid because it's so hot, but changes to a solid inner core because of the pressure.

Similar to the inner core, high pressures make the lower mantle a hard solid.

The crust and outermost mantle comprise the hard, solid lithosphere, which forms the rigid tectonic plates.

As temperature rises with increasing depth, the mantle softens to form the asthenosphere.

The structure of hard lithosphere, softer asthenosphere, and hard lower mantle make the outer portions of the Earth similar to an Oreo® cookie.

rising asthenosphere keeps the lower part of the crust hot—it takes cold, brittle material to make earthquakes.

At a **convergent plate boundary**, one tectonic plate dives ("subducts") beneath another and a line of volcanoes develops on the overriding plate. Earthquakes occur at a variety of depths as the cold, rigid plate ("lithosphere") plunges into the asthenosphere. The largest earthquakes on Earth occur where two converging plates lock together for centuries, then suddenly let go.

Shallow earthquakes, but little or no volcanic activity, occur where one plate slides past another at a **transform plate boundary**. And as a plate moves over a **hotspot**, magma rises from a plume of hot mantle material, causing shallow earthquakes and a line of volcanoes.

Most Earthquakes and Volcanic Eruptions Occur at Plate Boundaries and Hotspots

Volcanic eruptions and shallow earthquakes are common where plates rip apart.

Divergent Plate Boundary

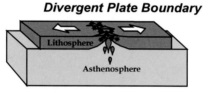

Robert J. Lillie

Where plates crash together, one dives ("subducts") beneath the other, causing volcanoes to erupt on the overriding plate and earthquakes at a variety of depths.

Convergent Plate Boundary

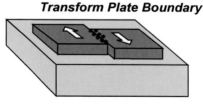

Robert J. Lillie

Shallow earthquakes and little volcanism occur where one plate slides laterally past another.

Transform Plate Boundary

Oreo® cookies are a fun way to demonstrate the three types of plate boundaries and a hotspot☺.

Robert J. Lillie

In places like Hawaii and Yellowstone, a plate rides over a rising plume of hot mantle, causing earthquakes and a chain of volcanoes.

Hotspot

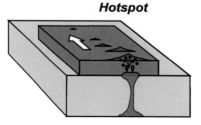

Robert J. Lillie

Volcanoes ◣ Earthquakes * Small to Moderate Size ☆ Very Large

"The Beasts" of Earthquakes and Volcanic Eruptions Strike Mostly at Plate Boundaries and Hotspots

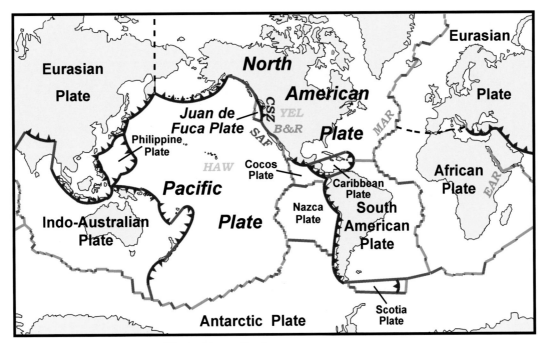

Plate Boundaries

Divergent \	Convergent (	Transform \	Hotspot
Divergent plate boundaries, such as the **Mid-Atlantic Ridge (MAR)**, **East African Rift (EAR)**, and **Basin and Range Province (B&R)**, produce volcanoes and shallow earthquakes.	Convergent plate boundaries have lots of volcanoes and most of the deeper earthquakes, as revealed by the Pacific "Ring of Fire" that includes the **Cascadia Subduction Zone (CSZ)**.	Transform plate boundaries, like the **San Andreas Fault (SAF)**, have shallow earthquakes and not much volcanism.	Earthquakes and volcanic activity also occur where plates move over hotspots like **Hawaii (HAW)** and **Yellowstone (YEL)**.

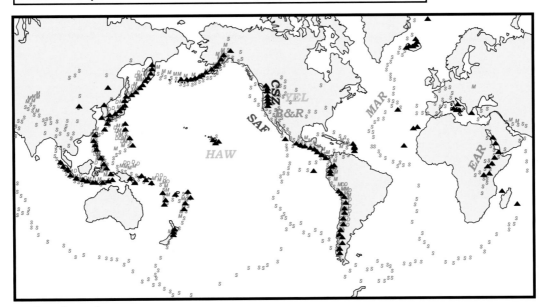

Earthquake Depths: S=Shallow (<40 miles)   M=Medium (40–200 Miles)   D=Deep (>200 Miles)

Volcanoes: ▲

The Greater Pacific Northwest has All Three Types of Plate Boundaries and a Hotspot

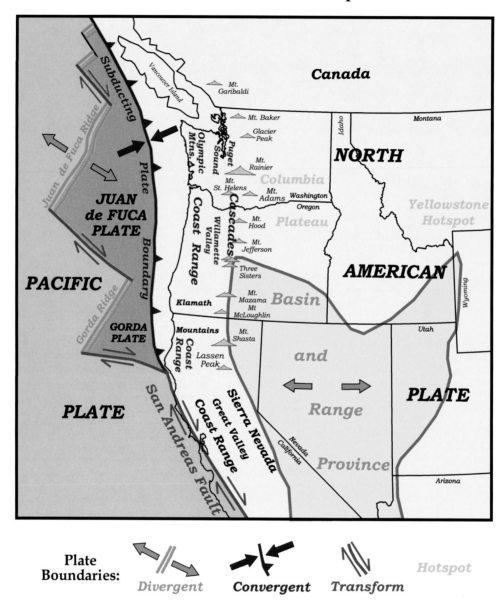

Plate Boundaries:

Divergent   Convergent   Transform   Hotspot

Subduction of the Juan de Fuca and Gorda plates beneath North America forms the coastal ranges and Cascade volcanoes. Offshore, the Juan de Fuca and Gorda plates diverge from the Pacific Plate at the Juan de Fuca and Gorda ridges. Between the ridge segments, the plates slide past one another along transform plate boundaries similar to California's San Andreas Fault. Onshore, the Basin and Range Province forms long mountain ranges and intervening valleys as plate divergence rips North America apart. The Columbia Plateau represents the surfacing of a hotspot now located beneath Yellowstone.

Landscapes of the Pacific Northwest and Surrounding Regions

Landforms and rocks of the greater Pacific Northwest reflect current or past plate-tectonic activity. All three types of plate boundaries—and a hotspot—are found in the region. The Basin and Range Province is a continental rift where plate divergence is starting to rip apart the North American Plate. The Coast Ranges and Cascades are parallel mountain ranges formed by the convergence of the Juan de Fuca and North American Plates along the Cascadia Subduction Zone. The California Coast Range and Sierra Nevada are similar parallel mountain ranges formed by earlier plate convergence. The San Andreas Fault disrupts this landscape as the Pacific Plate slides laterally past the North American Plate along a transform plate boundary. Volcanic materials of the Columbia Plateau and Yellowstone region have erupted—and continue to erupt— as the North American Plate moves in a westward direction over the Yellowstone Hotspot.

Plate tectonics helps us better understand the Northwest's varied ecology, natural plant and wildlife diversity, and human history. For example, the coastal ranges and Cascades create a "rain shadow" that causes storms coming from

National Parks in the Pacific Northwest and Surrounding Regions Display Landforms Developed at All Three Types of Plate Boundaries and a Hotspot

NP = National Park
NM = National Monument
NS = National Seashore
NRA = National Recreation Area
N Res = National Reserve
W-S-T = Whiskeytown-Shasta-Trinity

Other sites discussed in this book are managed by the U. S. Forest Service (USFS), Bureau of Land Management (BLM), and other federal and state agencies.

Tectonic Setting:

Divergent Plate Boundary (Basin and Range Province)

Convergent Plate Boundary (Cascadia Subduction Zone)

Transform Plate Boundary (San Andreas Fault)

Hotspot (Columbia Plateau – Yellowstone)

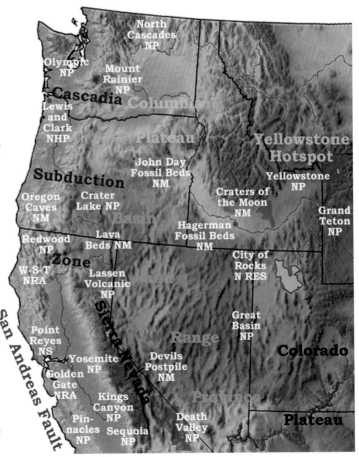

the Pacific Ocean to drop lots of rain and snow on western Oregon and Washington and very little east of the Cascades. The region's green western reaches and brown interior are thus products of mountain ranges forming where tectonic plates converge. But the varied climate zones, the result of slow processes acting over tens of millions of years, are being modified at accelerated rates by recent human activity.

Geological Hazards

The "Beast" of the Pacific Northwest and surrounding regions includes a variety of earthquakes and volcanic eruptions. These events are mostly due to the three types of plate boundaries and a hotspot that have built, and continue to build, the landscape.

Earthquakes

The largest concern for earthquakes in the Pacific Northwest is a "locked zone" earthquake, where the subducting Juan de Fuca Plate sticks to the North American Plate for centuries, then suddenly—and violently—lets go. These mega-earthquakes occur every 200 to 600 years or so, and the last one was in the year 1700. We are wise to prepare our homes, communities, and infrastructure for the next "Big One," and to know what to do during the ground shaking, landslides, and tsunamis that are sure to come.

Other types of earthquakes, particularly those that strike populated regions of Puget Sound in Washington and the Willamette Valley in Oregon, are also a concern. These include "slab earthquakes" that originate from the top portion of the subducting Juan de Fuca Plate, and "crustal earthquakes" along shallow fault lines cutting the North American Plate. "Volcanic earthquakes" can be associated with magma moving beneath the Cascades, Basin and Range Province, Juan de Fuca Ridge system, and Yellowstone Hotspot. The North American Plate rifting apart in the Basin and Range Province produces "rifting earthquakes," and "offshore earthquakes" are generated at the divergent and transform plate boundaries between the Pacific and Juan de Fuca Plates. And of course, lurking just to the south is the San Andreas Fault, a transform plate boundary that frequently causes small to moderate-sized earthquakes in western California. Knowledge of these earthquakes—and what to do to prepare for them—comes not only from scientific studies over the past century, but also from indigenous people who learned to take the "Beast with the Beauty" of the Northwest by passing knowledge from generation to generation.

Volcanoes

The landscapes, ecology, and cultural history of the greater Pacific Northwest are also tied to a variety of volcanic processes that have been occurring for millions of years. Volcanic activity continues to change the scenery and affect the region's environment and economy. Ever since Native Americans first arrived, people have been moved by the power and beauty of these volcanoes. The region showcases the full spectrum of volcanoes and their eruptive products.

Cascade volcanoes erupt directly above where the top of the subducting Juan de Fuca Plate reaches about 50 miles below the surface. Increased temperature and pressure at that depth cause the rocks to metamorphose and dehydrate ("sweat"). Via chemical reactions, the rising hot water causes overlying rock to melt, generating magma that at times erupts out on the surface. Other volcanic activity occurs offshore along the Juan de Fuca Ridge system, where the Juan de Fuca Plate diverges from the Pacific Plate. East of the Cascades, volcanic activity is associated with continental rifting in the Basin and Range Province. And much farther to the east, Yellowstone National Park lies above a hotspot that causes supervolcano eruptions.

Silica, composed of the elements silicon and oxygen, can be thought of as the "thickening agent" of molten magma—much like

Earthquakes and Volcanoes of the Pacific Northwest

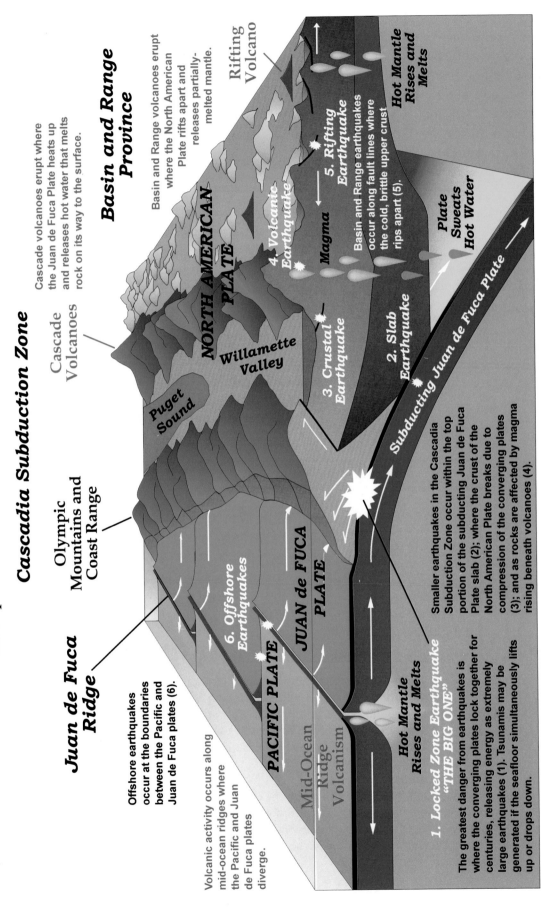

Cascadia Subduction Zone

Basin and Range Province

Cascade volcanoes erupt where the Juan de Fuca Plate heats up and releases hot water that melts rock on its way to the surface.

Basin and Range volcanoes erupt where the North American Plate rifts apart and releases partially-melted mantle.

Rifting Volcano

Hot Mantle Rises and Melts

5. Rifting Earthquake

Basin and Range earthquakes occur along fault lines where the cold, brittle upper crust rips apart (5).

Magma

4. Volcanic Earthquake

Plate Sweats Hot Water

NORTH AMERICAN PLATE

Willamette Valley

3. Crustal Earthquake

2. Slab Earthquake

Subducting Juan de Fuca Plate

Cascade Volcanoes

Puget Sound

Smaller earthquakes in the Cascadia Subduction Zone occur within the top portion of the subducting Juan de Fuca Plate slab (2); where the crust of the North American Plate breaks due to compression of the converging plates (3); and as rocks are affected by magma rising beneath volcanoes (4).

Olympic Mountains and Coast Range

Juan de Fuca Ridge

Volcanic activity occurs along mid-ocean ridges where the Pacific and Juan de Fuca plates diverge.

Offshore earthquakes occur at the boundaries between the Pacific and Juan de Fuca plates (6).

6. Offshore Earthquakes

PACIFIC PLATE

JUAN de FUCA PLATE

Mid-Ocean Ridge Volcanism

Hot Mantle Rises and Melts

1. Locked Zone Earthquake "THE BIG ONE"

The greatest danger from earthquakes is where the converging plates lock together for centuries, releasing energy as extremely large earthquakes (1). Tsunamis may be generated if the seafloor simultaneously lifts up or drops down.

flour in pancake batter. Eruptions generated by thick, pasty magma—rich in silica—produce lava flows that build volcanoes with steep slopes. Combined with volcanic ash, pumice, mudflows, and other materials, they build up to form steep cones known as composite volcanoes, such as Mt. Rainier, Mt. Adams, Mt. Hood, Mt. Jefferson, the Three Sisters, and Mt. Shasta in the Cascades. Portland's skyline reveals Mt. St. Helens, where pasty lava caused the violent 1980 eruption that blanketed the region with ash and filled waterways with slurries of mud. Since then, smaller volcanoes of sticky lava, known as lava domes, have been filling the crater of the exploded mountaintop.

Lava that is less pasty—relatively poor in silica—tends to flow for longer distances and form more gently-sloping volcanoes. Newberry Volcano in central Oregon and Medicine Lake Volcano in northern California are broad, low-profile shield volcanoes composed mostly of dark, free-flowing lava known as basalt. They are much like the big volcanoes that comprise the Hawaiian Islands. Numerous smaller volcanoes caused by gas-propelled eruptions of fluid basalt lava, known as cinder cones, dot the slopes of Newberry Volcano. Some familiar cinder cones in the Pacific Northwest include Wizard Island in Crater Lake; Pilot Butte and Lava Butte near Bend, Oregon; and Sconchin Butte in northern California's Lava Beds National Monument.

Classification of Igneous Rocks

		CHEMICAL COMPOSITION			
		Approximate % Silica (SiO$_2$)			
		70%	60%	50%	40%
TEXTURE	Fine-Grained Extrusive ("Volcanic")	Rhyolite	Andesite	Basalt	
	Coarse-Grained Intrusive ("Plutonic")	Granite	Diorite	Gabbro	Peridotite

Igneous rocks form when molten Earth material (magma) cools. Their general classification depends on the size of the mineral grains (texture) and the types of minerals found within the rock (chemical composition).

1. <u>Texture</u>. Where cooling is slow (deep below the surface) mineral crystals have time to grow large, forming intrusive (plutonic) rocks. When magma flows out on the surface as lava it cools before large crystals can grow, forming extrusive (volcanic) rocks.

2. <u>Chemical Composition</u>. High-silica rocks cool from pasty magma that forms intrusive granite below the surface and extrusive rhyolite above. Magmas with lesser amounts of silica tend to flow more freely, forming grabbro at depth and basalt on the surface. Earth's mantle is a very low-silica rock known as peridotite.

The Amount of Silica in Magma Determines the Type and Shape of Volcano

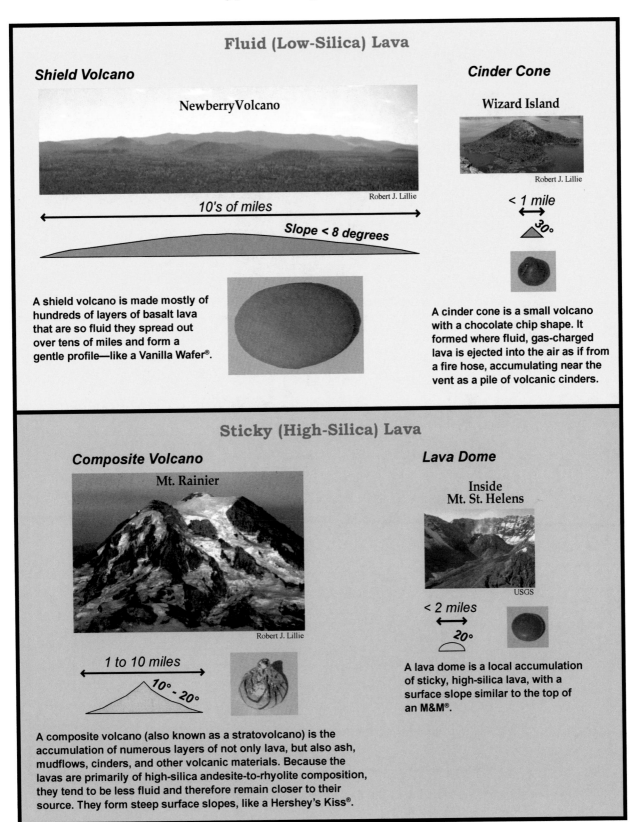

Fluid (Low-Silica) Lava

Shield Volcano

NewberryVolcano

Robert J. Lillie

10's of miles

Slope < 8 degrees

A shield volcano is made mostly of hundreds of layers of basalt lava that are so fluid they spread out over tens of miles and form a gentle profile—like a Vanilla Wafer®.

Cinder Cone

Wizard Island

Robert J. Lillie

< 1 mile

30°

A cinder cone is a small volcano with a chocolate chip shape. It formed where fluid, gas-charged lava is ejected into the air as if from a fire hose, accumulating near the vent as a pile of volcanic cinders.

Sticky (High-Silica) Lava

Composite Volcano

Mt. Rainier

Robert J. Lillie

1 to 10 miles

10° - 20°

A composite volcano (also known as a stratovolcano) is the accumulation of numerous layers of not only lava, but also ash, mudflows, cinders, and other volcanic materials. Because the lavas are primarily of high-silica andesite-to-rhyolite composition, they tend to be less fluid and therefore remain closer to their source. They form steep surface slopes, like a Hershey's Kiss®.

Lava Dome

Inside Mt. St. Helens

USGS

< 2 miles

20°

A lava dome is a local accumulation of sticky, high-silica lava, with a surface slope similar to the top of an M&M®.

The volcanic landscapes in the Pacific Northwest preserve not only a wide variety of materials erupted from volcanoes, but also the impressions of a dynamic Earth that volcanic events leave on the minds of human observers. Northwest native people, for example, witnessed eruptions of steam, ash, mudflows, and possibly lava flows from Mount Rainier during the past 5,600 years. Their stories of geologic events, while woven with mythic lore, are vivid in description. Their oral traditions tell of "blood running down the slopes" (lava flows); the mountain losing its top and the ensuing landslide and mudslide wiping trees off valley slopes and filling the Puyallup Valley with bubble-filled stones; and the Kent-Auburn Valley as an estuary that has since been filled in with sediment that originated from Mount Rainier mudflows.

Building the Northwest Landscape

Most of the Pacific Northwest did not even exist 200 million years ago. The Pacific coastline at that time ran through what is now Idaho. Gradually, like groceries piling up on the conveyor belt in a supermarket checkout line, various pieces were added to the edge of the continent. Together with material erupted from

Growth of the Pacific Nothwest

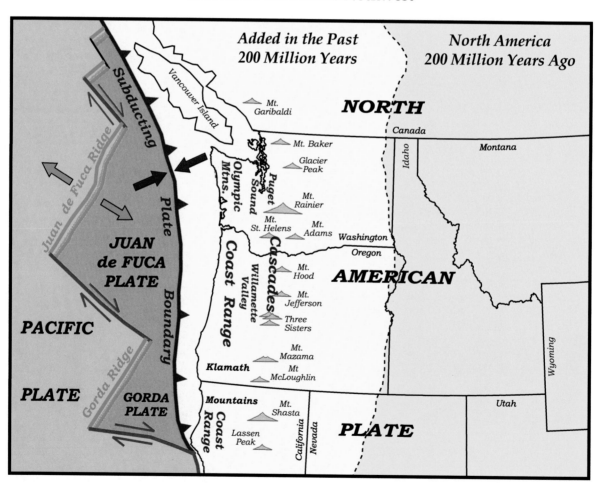

200 million years ago the coastline of the Pacific Northwest was near Idaho's western border. There was no Oregon and Washington!

volcanoes and marine sedimentary and volcanic layers scraped off the ocean floor, these added pieces have gradually formed the landscapes of Washington, Oregon and northern California so familiar to us today. The Pacific Northwest is still growing, as material is added along the coast and in the Cascades.

When viewed over millions of years, the growth of a continent and development of its landscape may seem gradual and continuous. But much of the growth is the more discrete work of the Beast. Individual earthquakes can offset or lift the land a fraction of an inch or a few inches at a time. That may not seem like much, but when tens of thousands of earthquakes occur over a few million years, the Olympic

Mountains and other coastal ranges form. Likewise, individual volcanic eruptions may add only a thin layer to the surface. But thousands of eruptions over half a million years have built Mt. Rainier to a 14,000-foot volcanic peak.

The spectacular landscapes of the Pacific Northwest are truly "Beauty from the Beast." An individual beast—an earthquake or a volcanic eruption—may temporarily cause pain and destruction. But the cumulative beast—ground movement or the accumulation of volcanic material—can raise the Teton Mountains, build majestic Mt. Rainier, carve out the landscape of San Francisco Bay, and produce the geysers, hot springs, mud pots, and other geothermal features of Yellowstone.

The Pacific Northwest has Grown Westward over the past 200 Million Years by Subduction and Terrane Accretion

The conveyer belt represents the oceanic plate subducting beneath the continental plate.

Robert J. Lillie

A grocery store checkout counter can be used to demonstrate the gradual growth of a continent via terrane accretion.

Robert J. Lillie

Robert J. Lillie

Robert J. Lillie

Robert J. Lillie

The groceries are like volcanic islands and continental fragments ("terranes") that progressively crash into the edge of the continent, adding to its mass.

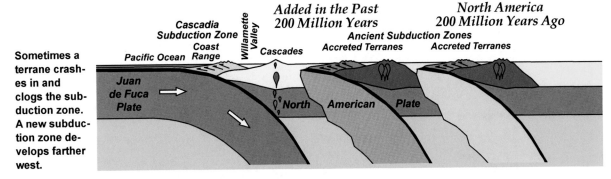

The Cascadia Subduction Zone is just the latest of several subduction zones that developed in the western United States over the past 200 million years.

Beauty from the Beast: Landscapes

Divergent Plate Boundary

Basin and Range topography results from movement along faults during tens of thousands of earthquakes over a few million years.

Transform Plate Boundary

Narrow valleys and long ridges developed as the San Francisco Bay Area was sheared-up during tens of thousands of earthquakes along the San Andreas Fault during the past 20 million years.

Convergent Plate Boundary

Mt. Rainier is a composite volcano formed during thousands of eruptions of lava, ash, cinders, and mudflows over the past half a million years.

Hotspot

Spectacular geysers, hot springs, and other geothermal features form on the crest of a supervolcano.

Additional Reading

Alt, David D. and Donald W. Hyndman, 1995, "Northwest Exposure: A Geologic Story of the Northwest," Missoula, MT: Mountain Press Publishing Company, 443 pp.

Alt, David D. and Donald W. Hyndman, 1978, "Roadside Geology of Oregon," Missoula, Mt: Mountain Press Publishing Company, 278 pp.

Alt, David D. and Donald W. Hyndman, 1984, "Roadside Geology of Washington," Missoula, MT: Mountain Press Publishing Company, 288 pp.

Alt, David D. and Donald W. Hyndman, 1989, "Roadside Geology of Idaho," Missoula, MT: Mountain Press Publishing Company, 403 pp.

Alt, David D. and Donald W. Hyndman, 2000, "Roadside Geology of Northern and Central California," Missoula, MT: Mountain Press Publishing Company, 384 pp.

Bishop, Ellen M., 2014, "Living with Thunder: Exploring the Geologic Past, Present, and Future of the Pacific Northwest," Corvallis, OR: OSU Press, 160 pp.

Brown, Cynthia L., 2011, "Geology of the Pacific Northwest: Investigating how the Earth was Formed," White River Junction, VT: Nomad Press, 120 pp.

Chronic, H., 1986, "Pages of Stone: Geology of Western National Parks and Monuments (Sierra Nevada, Cascades and Pacific Coast, Vol. 2)," Seattle, WA: Mountaineers Books, 184 pp.

Condie, K. C., 1982, "Plate Tectonics and Crustal Evolution," 2nd ed., New York: Pergamon Press, 310 pp.

Coney, P., D. Jones, and J. Monger, 1980, "Cordilleran suspect terranes," Nature, v. 288, p. 329-333.

Decker, R., and B. Decker, 2001, "Volcanoes in America's National Parks," New York: W. W. Norton and Comp., 256 pp.

Duffield, Wendell A., 2011, "What's so Hot about Volcanoes?" Missoula, MT: Mountain Press Publishing Company, 90 pp.

Harris, A. G., E. Tuttle, and S. P. Tuttle, 2004, "Geology of National Parks," 6th Ed., Dubuque, IA: Kendall/Hunt Pub. Comp., 882 pp.

Howell, D. G., 1995, "Principles of Terrane Analysis: New Applications for Global Tectonics," 2nd ed., New York: Chapman and Hall, 245 pp.

Jones, D. L., and others, 1982, "The growth of western North America," Scientific American, v. 247, p. 70-84.

Lillie, Robert J., 1999, "Whole Earth Geophysics: An Introductory Textbook for Geologists and Geophysicists," Upper Saddle River, New Jersey: Prentice Hall, Inc., 361 pp.

Lillie, Robert J., 2005, "Parks and Plates: The Geology of Our National Parks, Monuments, and Seashores," New York: W. W. Norton and Company, 298 pp.

Lillie, R. J., A. Mathis, and R. Riolo, 2011, "Geology: A Living Stage of Our Past, Present, and Future," Legacy, National Association for Interpretation, p. 8-11, January/February.

McPhee, J., 1998, "Annals of the Former World," New York: Farrar, Straus and Girous. 696 pp.

Miller, Marli B., 2014, "Roadside Geology of Oregon," 2nd Ed, Missoula, Mt: Mountain Press Publishing Company, 380 pp.

Moores, E. M. (editor), 1990, "Shaping the Earth: Tectonics of Continents and Oceans," Readings from Scientific American, New York: W. H. Freeman and Comp., 206 pp.

Orr, W. N., and E. L. Orr, 2002, "Geology of the Pacific Northwest," 2nd Edition, Long Grove, IL: Waveland Press, 337 pp.

Spearing, Darwin, and David Lageson, 1988, "Roadside Geology of Wyoming," Missoula, MT: Mountain Press Publishing Company, 288 pp.

Williams, Hill, 2002, "The Restless Northwest: A Geological Story," Pullman, WA: Washington State University Press, 176 pp.

Windley, B. F., 1995, "The Evolving Continents," 3rd ed., New York: John Wiley and Sons.

1

Divergent Plate Boundary: Basin and Range Province

The Basin and Range Province is a region where the North American continent is ripping apart—a divergent plate boundary in the making. It includes all of Nevada and portions of Wyoming, Utah, California, and Arizona, and extends into southern Oregon, Idaho, and Montana. The Basin and Range Province is similar to the East African Rift, where volcanic materials and sedimentary layers deposited by rivers and lakes partially fill rift valleys. Prominent planar surfaces (fault escarpments) along range fronts are evidence that the region continues to rip apart—erosion would smooth them out if the fault movement stopped.

The Basin and Range Province has Long Mountain Ranges Separated by Broad Valleys

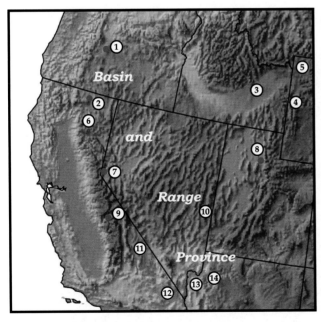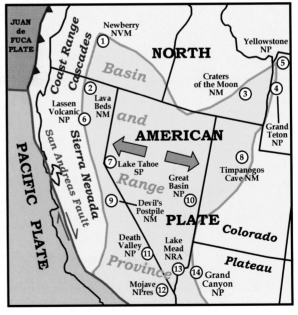

Areas in red highlighted by numbers are administered by the National Park Service, except for Newberry National Volcanic Monument, a U. S. Forest Service site in Oregon (1), and Lake Tahoe State Park in Nevada (7). NP = National Park; NM = National Monument; NPres = National Preserve; NRA = National Recreation Area; NVM = National Volcanic Monument; SP = State Park.

← Lake Tahoe, California/Nevada. Deep lakes form when a continent rips apart. (Photo by Robert J. Lillie).

The same continental rifting processes that form the high-elevation valleys and mountain ranges of the Basin and Range Province also result in earthquakes and volcanic activity. As the plate of lithosphere pulls apart and thins, hot asthenosphere rises and expands, elevating the entire region. The drop in pressure on the shallow asthenosphere makes it start to melt. Some of the liquid rock (magma) makes it to the surface and forms lava flows, shield volcanoes, cinder cones, and other volcanic features. As forces try to pull it apart, the cold, brittle upper crust breaks along fault lines, causing shallow earthquakes like the two that occurred beneath Oregon's Klamath Basin in 1993. Prolonged stretching and thousands of earthquakes result in down-dropped valleys (basins) separating uplifted mountains (ranges).

Ripping a Continent Apart

Earth's crust beneath continents is thicker than the crust underlying oceans. As a plate capped by thick continental crust rips apart, the mantle portion thins. The underlying hot mantle ("asthenosphere") flows upward to fill the void. Being under less pressure, the shallow asthenosphere expands—much like inflating a hot-air balloon—and lifts up the topography. We can see this in the Basin and Range Province.

Divergent Plate Boundary Development

Plate with Thick Continental Crust

Continental crust is thick and buoyant, and therefore sticks up above sea level.

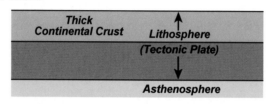

Plate Rips Apart

As the plate stretches and thins, the underlying asthenosphere flows upward and expands like a hot-air balloon, lifting the region to higher elevations.

The continental crust breaks along faults, forming long mountain ranges separated by rift valleys.

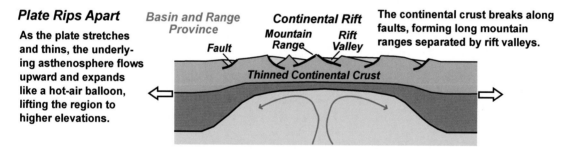

Ocean Basin Opens

If divergence continues, the continental crust completely breaks apart and thinner oceanic crust forms between the two continental blocks.

The ocean basin sinks below sea level because the crust is thinner and more dense, and therefore less buoyant.

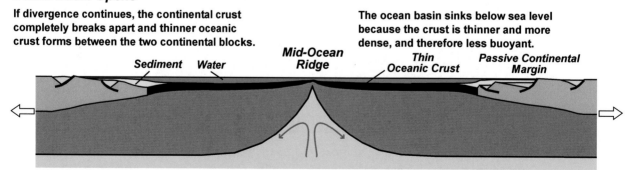

The floors of rift valleys, like the Klamath Basin in Oregon and Jackson Hole in Wyoming, are more than 4,000 feet above sea level. Adjacent mountain ranges are even higher—for example, Steens Mountain in southeastern Oregon is over 9,000 feet elevation and the Teton Range in Wyoming rises to more than 13,000 feet.

Lower pressure on the asthenosphere also has another important effect. At its normal depths beneath the lithosphere, this part of Earth's mantle is solid because it is under so much pressure. But if it rises fast enough it remains hot. Under the lowered pressure it begins to melt (much like superheated water flashes to steam when the lid is suddenly removed from a pressure cooker). Molten rock (magma) thus melts off the decompressed mantle at a continental rift. Some of it rises and invades the overlying crust to form a dense intrusive rock known a gabbro (the slow-cooling, course-grained equivalent of basalt). In some places the magma reaches the surface and spews out as lava flows and other volcanic products.

The lower crust in a continental rift generally tends to flow, like hot plastic, because it has been heated up by the shallow asthenosphere. But the upper part of the crust remains relatively cold. Like peanut brittle, it tends to crack along fault lines, causing earthquakes. Similar to a falling row of dominoes, blocks of the upper crust rotate and move apart, forming the long valleys (basins) separated by mountains (ranges) so typical of the Basin and Range Province and other continental rifts.

If plate divergence continues, a continent completely rips apart and the blocks of thick crust move away from one another. Volcanic processes continue to manufacture crust, but it is the heavy (basalt/gabbro) variety and is much thinner than the crust underlying the continents. The region sinks below sea level and a small ocean, like the Red Sea, forms. With continued plate divergence the separated continents move away from one another and a broad ocean, like the Atlantic, develops. This suggests an interesting real-estate opportunity for Northwest residents. For a (very) long-term investment, they might consider purchasing "ocean front" property in what is now Las Vegas, Nevada or Klamath Falls, Oregon☺.

Continental Rift: Topography, Earthquakes and Volcanism

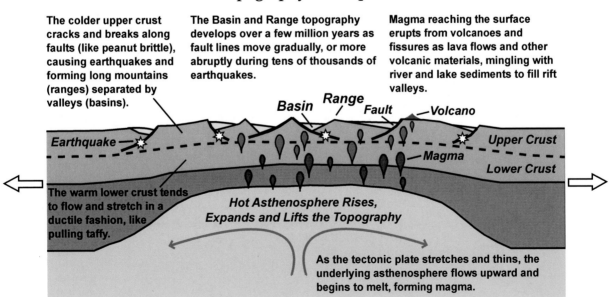

The colder upper crust cracks and breaks along faults (like peanut brittle), causing earthquakes and forming long mountains (ranges) separated by valleys (basins).

The Basin and Range topography develops over a few million years as fault lines move gradually, or more abruptly during tens of thousands of earthquakes.

Magma reaching the surface erupts from volcanoes and fissures as lava flows and other volcanic materials, mingling with river and lake sediments to fill rift valleys.

The warm lower crust tends to flow and stretch in a ductile fashion, like pulling taffy.

Hot Asthenosphere Rises, Expands and Lifts the Topography

As the tectonic plate stretches and thins, the underlying asthenosphere flows upward and begins to melt, forming magma.

Basin and Range Landscape

The topography of the Basin and Range Province reveals the full range of characteristics of a continental rift. First, much of the region—particularly the northern portion—is well above sea level. It's as if the whole landscape is rising upward, with the ranges going up a little faster than the adjacent basins. For example the floor of a basin in Wyoming, Jackson Hole, is 6,000 feet above sea level, while the adjacent Teton Range rises to over 13,000 feet.

Beauty from the Beast at a Continental Rift

Grand Teton National Park, Wyoming

Teton Range

Jackson Hole

Snake River

National Park Service

The east side of the Teton Range is a steep fault escarpment rising from the adjacent basin (Jackson Hole).

Lava Beds National Monument, California

Gillem Bluff is a small range with a steep fault escarpment.

Gillem Bluff

Fault Escarpment

Tule Lake

Lava Flow

Sediments

Robert J. Lillie

Tule Lake basin is partially filled with sediments and lava flows.

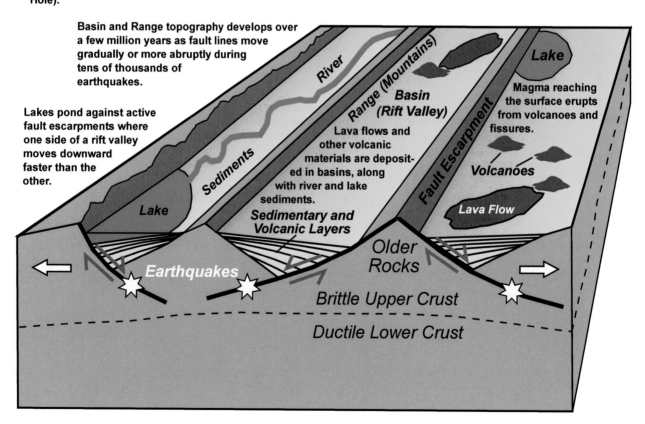

Basin and Range topography develops over a few million years as fault lines move gradually or more abruptly during tens of thousands of earthquakes.

Lakes pond against active fault escarpments where one side of a rift valley moves downward faster than the other.

River

Range (Mountains)

Basin (Rift Valley)

Lava flows and other volcanic materials are deposited in basins, along with river and lake sediments.

Sedimentary and Volcanic Layers

Lake

Magma reaching the surface erupts from volcanoes and fissures.

Volcanoes

Lava Flow

Fault Escarpment

Sediments

Lake

Earthquakes

Older Rocks

Brittle Upper Crust

Ductile Lower Crust

Deep Lakes and Deep Valleys form at Continental Rifts

Lake Tahoe, California/Nevada

Robert J. Lillie

At 1,645 feet deep, Lake Tahoe is the world's 8th deepest lake.

Death Valley National Park, California/Nevada

Elevation: –282 Feet

Badwater Basin

Robert J. Lillie

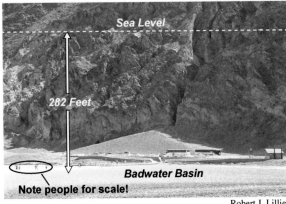

Sea Level

282 Feet

Badwater Basin

Note people for scale!

Robert J. Lillie

At 282 feet below sea level, Badwater Basin is the lowest point in North America. During the last ice age, much of Death Valley was filled with Lake Manly—about 80 miles long and 600 feet deep.

But in the southern part of the province, where rifting is more advanced, elevations are generally lower—much of the floor of Death Valley in eastern California is below sea level!

Layers of sand, mud and gravel deposited by rivers and lakes, along with lava flows and other volcanic materials, fill rift valleys as they form. But sometimes the valley floors move downward much faster than these layers can fill them. Thus a second characteristic of continental rifts is that their valleys contain most of the deepest lakes in the world. These include the world's deepest, Lake Baikal in Siberia (5,387 feet deep) and the 2nd and 4th deepest, Lake Tanganyika (4,323 feet) and Lake Malawi (2,316 feet), in the East African Rift. The Basin and Range Province has the world's 8th deepest lake, Lake Tahoe on the California/Nevada border (1,645 feet).

During the waning stages of the latest ice age, between 20,000 and 10,000 years ago, the Basin and Range Province was much cooler and wetter than it is today. Basins thus held much more water than they do today. For example, Death Valley contained Lake Manly, some 80 miles long and 600 feet deep. Such lakes were important to early Americans. Archaeological sites along ancient shorelines—today more than 300 feet above the levels of modern "pluvial" lakes—reveal much about these earlier cultures.

The Beast of Rifting

Twenty million years ago, much of what is now the Basin and Range Province was relatively flat, oak-studded grassland. Then the crust began to stretch apart. That process continues today. Consequences include breaks (faults) in the crust, as well as volcanic eruptions. Faults caused mountain ranges, such as Steens Mountain and Abert Rim in southeastern Oregon, to rise above adjacent basins (the Alvord Desert and Abert Lake). When these fault lines move suddenly, small-to-medium-sized earthquakes shake the region. The continental rifting has also resulted in a vast amount of volcanic activity, including Oregon's most recent eruption, the Big Obsidian Flow inside the caldera of Newberry Volcano 1,300 years ago, and eruptions of Medicine Lake Volcano in California 1,100 years ago.

Earthquakes

Being pulled apart is pretty stressful. In fact, the Earth cracks. Large blocks of the crust move up and down along the cracks, known as faults. A series of north-south oriented mountains (ranges), separated by valleys (basins), is created. This pattern of basins and ranges repeats itself over and over. In southeastern Oregon, for example, one sees the Alvord Desert (basin) and

Steens Mountain (range); Warner Valley (basin) and Hart Mountain (range); Lake Abert (basin) and Abert Rim (range); Summer Lake (basin) and Winter Ridge (range). Thousands and thousands of earthquakes over the past 20 million years—some large, some small—produced the basins and ranges we see today. And the Earth continues to shake.

Notable Basin and Range earthquakes include two that occurred in Oregon's Klamath Basin in 1993, and two that occurred on the northeast fringes of the province. The August 17, 1959 Hebgen Lake earthquake in southwest Montana, at magnitude 7.5, was the second largest to strike the lower 48 states in the 20th Century (the largest was the 1906 San Francisco

Earthquakes Rattle the Basin and Range Province

Klamath Falls, Oregon

Lou Sennick, *Herald and News*, Klamath Falls, Oregon

On September 20, 1993, two magnitude 6 earthquakes occurred along a fault line bounding the west side of the Klamath Basin.

earthquake along the San Andreas Fault). The shaking caused a landslide that killed 28 people camped along the Madison River, and ponded the river to form Earthquake Lake. On October 28, 1983, a magnitude 6.9 earthquake occurred near Idaho's highest mountain, Borah Peak. The earthquake killed two children walking to school in nearby Challis, Idaho. The fault movement offset the ground surface, forming a 10-foot high escarpment that extended over a 20-mile length at the base of the Lost River Range.

Volcanism

The magma that comes from the partially-melted mantle in a continental rift zone has low silica (basalt) composition. Indeed, a large amount of the lava flows in the Basin and Range Province form the heavy, dark-colored basalt lava-rock characteristic of the region. When hundreds of fluid basalt lava flows pile up over tens of thousands of years, broad shield volcanoes form. Prominent examples in the Basin and Range Province include Newberry Volcano

Earthquakes Build Landscapes Over Time

Borah Peak, Idaho

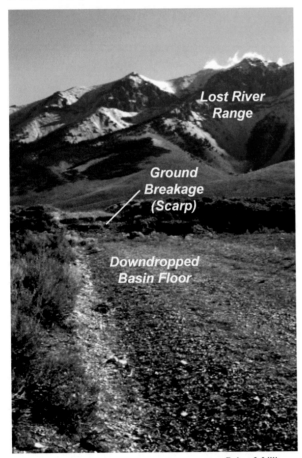

Robert J. Lillie

Movement along the fault line caused the basin floor to drop down up to 10 feet relative to the adjacent mountain range.

Robert J. Lillie

A fresh scarp in the landscape developed as the basin floor dropped downward during the earthquake.

Robert J. Lillie

View looking back across the downdropped scarp. During tens of thousands of earthquakes occurring over a few million years, miles of relief develops between a mountain range and an adjacent rift basin.

A magnitude 6.9 earthquake struck the Lost River Range near Borah Peak, Idaho on October 28, 1983.

Basin and Range Volcanism

Low-Silica Volcanism

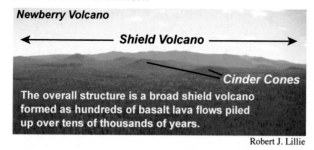

Newberry Volcano

← **Shield Volcano** →

Cinder Cones

The overall structure is a broad shield volcano formed as hundreds of basalt lava flows piled up over tens of thousands of years.

Robert J. Lillie

Basalt lava also formed over 400 cinder cones on the flanks of the volcano.

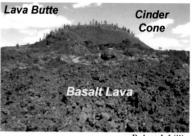

Lava Butte   **Cinder Cone**

Basalt Lava

Robert J. Lillie

High-Silica Volcanism

Summit Caldera

Central Pumice Cone   **Big Obsidian Flow**   **East Lake**

Paulina Lake

Robert J. Lillie

The summit caldera is partially filled with high-silica pumice and obsidian, as well as two large lakes.

Big Obsidian Flow

Obsidian

Robert J. Lillie

Newberry Volcano in Newberry National Volcanic Monument, Oregon, reveals both low-silica (basalt) and high-silica (rhyolite) volcanism. This pattern is characteristic of the early stages of volcanic activity in continental rifts, such as the Basin and Range Province.

Cinder Cones Form where Fluid, Gas-Charged Lava Erupts like a Fire Hose

Particles of solidified lava, including large volcanic bombs and smaller cinders, rain down around the vent and form a cone.

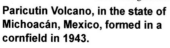

U. S. Geological Survey

Paricutin Volcano, in the state of Michoacán, Mexico, formed in a cornfield in 1943.

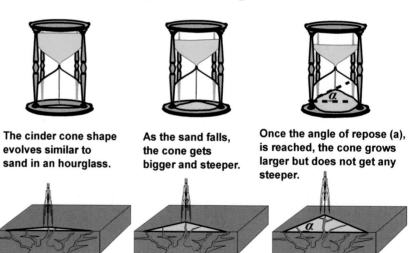

The cinder cone shape evolves similar to sand in an hourglass.

As the sand falls, the cone gets bigger and steeper.

Once the angle of repose (a), is reached, the cone grows larger but does not get any steeper.

south of Bend, Oregon, and Medicine Lake Volcano in northeastern California. When propelled by escaping gases, the fluid basalt lava often rains down as volcanic cinders, forming the hundreds of cinder cones scattered throughout the region.

In some places, though, the basalt magma does not make it cleanly to the surface. Rather, it melts its way through the thick continental crust. This tends to increase the amount of silica, which is the thickening agent for magma. A more pasty magma with a composition closer to that of granite forms. Where the magma reaches the surface it cools quickly and does not have time for the large crystal grains characteristic of granite to form. The lava flows formed are known as rhyolite, which is lighter both in color and weight than basalt. High-silica rhyolite composition can also form frothy, very light-weight pumice—a rock that can float on water! If the lava cools so quickly that it does not even have time for crystals to form, a glassy rock known as obsidian can form.

As with most of the natural world, many of Earth's features do not fit neatly into one simple explanation. Newberry Volcano in Oregon and Medicine Lake Volcano in California lie at the juncture of the Basin and Range Province and the Cascades Volcanoes. It is quite likely that volcanic activity in the region is influenced by both continental rifting on the east and plate subduction from the west. Likewise, portions

Fluid, Basalt Lava forms Numerous Cinder Cones in National, State and other Parklands in the Basin and Range Province

Lava Beds National Monument, California

Schonchen Butte

Robert J. Lillie

The top of Sconchin Butte is a popular hiking destination along the main road through Lava Beds National Monument.

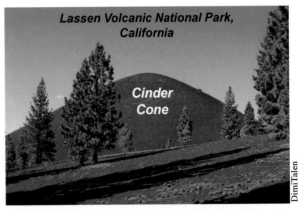

Lassen Volcanic National Park, California

Cinder Cone

DimiTalen

The eastern part of Lassen Volcanic National Park displays basaltic volcanisim characteristic of continental rifting in the Basin and Range Province (including a cinder cone cleverly named "Cinder Cone" ☺).

Craters of the Moon National Monument, Idaho

Broken Top Volcano

Robert J. Lillie

Craters of the Moon National Monument lies along the Columbia Plateau–Yellowstone hotspot track, but recent volcanism is also due to Basin and Range continental rifting.

of Pacific Northwest parks lie in zones of transition between different tectonic provinces. For example, the western half of Lassen Volcanic National Park in northern California has composite volcanoes and high-silica lava domes characteristic of the Cascadia Subduction Zone, while cinder cones and basalt lava flows in the eastern part of the park are more diagnostic of continental rifting in the Basin and Range Province. And while Craters of the Moon National Monument in southern Idaho lies along the Columbia Plateau–Yellowstone hotspot track, recent basalt fissure eruptions and cinder cones are most likely due to Basin and Range continental rifting.

Additional Reading

Alt, David D. and Donald W. Hyndman, 1978, "Roadside Geology of Oregon," Missoula, Mt: Mountain Press Publishing Company, 278 pp.

Alt, David D. and Donald W. Hyndman, 1989, "Roadside Geology of Idaho," Missoula, MT: Mountain Press Publishing Company, 403 pp.

Alt, David D. and Donald W. Hyndman, 2000, "Roadside Geology of Northern and Central California," Missoula, MT: Mountain Press Publishing Company, 384 pp.

Bishop, Ellen M., 2003, "In Search of Ancient Oregon," Portland, OR: Timber Press, 288 pp.

Bishop, Ellen M., 2014, "Living with Thunder: Exploring the Geologic Past, Present, and Future of the Pacific Northwest," Corvallis, OR: OSU Press, 160 pp.

Collier, M., 1990, "An Introduction to the Geology of Death Valley," Death Valley, California: Death Valley Natural History Association, 60 pp.

Decker, R., and B. Decker, 2001, "Volcanoes in America's National Parks," New York: W. W. Norton and Comp., 256 pp.

Good, J. M., and K. L. Pierce, 1996, "Interpreting the Landscapes of Grand Teton and Yellowstone National Parks," Moose, Wyoming: Grand Teton Natural History Assoc., 58 pp.

Lamb, Susan, 1991, "Lava Beds National Monument," Tulelake, CA: Lava Beds Natural History Association, 48 pp.

Love, J. David, John C. Reed, and Kenneth L. Price, 2007, "Creation of the Teton Landscape: A Geological Chronicle of Jackson Hole and the Teton Range," Moose, WY: Grand Teton Association, 120 pp.

Love, J. D., J. C. Reed, Jr., R. L. Christianson, and J. R. Stacy, 1973, "Geologic Block Diagram and Tectonic History of the Teton Range, Wyoming–Idaho," Washington, D.C.: U. S. Geological Survey, Misc. Geologic Invest., Map I-730.

McPhee, John, "Basin and Range," 1981, New York, NY: Noonday Press–Farrar, Straus, Giroux, 216 pp.

Miller, Marli B., and Lauren A. Wright, 2004, "Geology of Death Valley National Park," 2nd Ed, Death Valley, CA: Death Valley Natural History Association, 128 pp.

Miller, Marli B., 2014, "Roadside Geology of Oregon," 2nd Ed, Missoula, Mt: Mountain Press Publishing Company, 380 pp.

Orr, Elizabeth L. and William N. Orr, 2000, "Oregon Geology," 6th Edition, Corvallis, OR: Oregon State University Press, 304 pp.

Spearing, Darwin, and David Lageson, 1988, "Roadside Geology of Wyoming," Missoula, MT: Mountain Press Publishing Company, 288 pp.

2

Convergent Plate Boundary: Cascadia Subduction Zone

The landscape of the Pacific Northwest greatly influences the region's ecology, history, and commerce. Most Northwest residents live in a fertile lowland between coastal mountains and active volcanoes. Native Americans thrived on natural resources from the sea in the Puget Sound region of Washington and from land and river in the Willamette Valley of Oregon. When European settlers came from the east they generally bypassed the dryer regions of eastern Oregon and Washington to establish farms, fisheries, and other livelihoods in this low region between the mountains.

Subduction in the Pacific Northwest creates not only a very livable situation, but also incredible beauty in the form of coastlines, mountains, and valleys. This beauty is part of the reason that residents and visitors have learned to tolerate, and develop ways to mitigate, the effects of earthquakes, tsunamis, volcanic eruptions, and landslides—beasts that are consequences of the same tectonic forces that create the beauty and livability.

Views of the Cascadia Subduction Zone

Coast Range

Marys Peak

Robert J. Lillie

Cascade Volcanoes

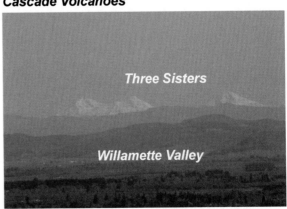

Three Sisters

Willamette Valley

Robert J. Lillie

Fitton Green, on the eastern edge of the Oregon Coast Range near Corvallis, offers a panoramic view of the Cascadia Subduction Zone.

Cascadia Landscape

North America has been building westward over the past 200 million years by subduction. As an oceanic plate subducts, it adds material to the edge of the continent as layers scrape off the ocean floor, volcanoes erupt, and islands and continental fragments slam into the coast. The current Cascadia Subduction Zone is just the most recent subduction zone. Its landscape features and geologic activity profoundly influence the Pacific Northwest's history, climate, safety—and beauty!

Two parallel mountain ranges have been forming in the Pacific Northwest as a result of the Juan de Fuca Plate subducting beneath the edge of North America. The Coast Range consists of sedimentary rock layers and hard crust scraped off the ocean floor where the plate begins to dive downward. The grinding action also produces devastating earthquakes, including some that result in giant tsunami waves.

Beauty from the Beast: Cascadia Subduction Zone

Coast Range

Robert J. Lillie

The picturesque Oregon Coast forms where oceanic layers are scraped off the subducting Juan de Fuca Plate and plastered to the edge of North America.

Willamette Valley

Robert J. Lillie

The rich farmland of the Willamette Valley lies near sea level between the Coast Range and Cascade Volcanoes.

Cascade Volcanoes

Robert J. Lillie

The scenery of the Cascade Volcanoes belies their firey origins.

Robert J. Lillie

The beauty of Bumpass Hell is due to colorful mineral deposits from hydrothermal features in an active volcanic range.

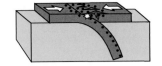

The Cascades are a volcanic mountain range extending from southern British Columbia, through Washington and Oregon, and into northern California. As the Juan de Fuca Plate descends deep into the Earth it gets hotter and starts to dehydrate—that is, it "sweats" hot water. The rising water causes some of the rock of the overriding North American Plate to melt, generating molten rock, known as magma. Some of the magma erupts on the surface as lava flows and other materials forming Mt.

Rainier, Mt. Hood, Mt. Shasta, and other Cascade volcanoes.

These two mountain ranges influence many of the physical and cultural aspects of the Pacific Northwest. Storms coming in from the Pacific Ocean drop most of their moisture on the Coast Range and Cascades, leaving central and eastern Oregon high and dry. This affects not only plant and animal communities, but also human habitation and land use. The physical landscape was an important component of the

Beast of the Cascadia Subduction Zone

Coast

Robert J. Lillie

Ancient stumps are the remains of trees thought to have been killed by sea water when the costal region suddenly dropped during a massive Cascadia Subduction Zone earthquake.

Western Cascades

KOMO News

A landslide in 2014 devasted Oso, Washington, in the over-steepened terrain of the Western Cascades.

Cascade Volcanoes

Robert J. Lillie

Lava flows from Little Belknap Crater 1500 years ago covered hundreds of square miles.

Robert J. Lillie

Lassen Peak erupted violently from 1914 to 1917.

29

Plate Convergence Leads to Subduction and the Formation of Two Parallel Mountain Ranges

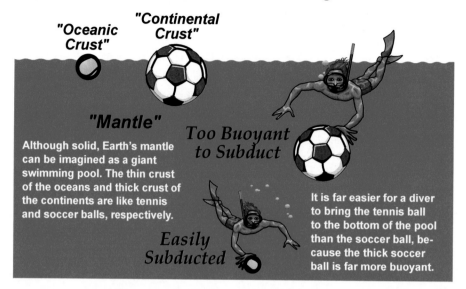

"Oceanic Crust"

"Continental Crust"

"Mantle"

Although solid, Earth's mantle can be imagined as a giant swimming pool. The thin crust of the oceans and thick crust of the continents are like tennis and soccer balls, respectively.

Too Buoyant to Subduct

Easily Subducted

It is far easier for a diver to bring the tennis ball to the bottom of the pool than the soccer ball, because the thick soccer ball is far more buoyant.

Where tectonic plates converge, the one capped by the thin oceanic crust dives ("subducts") beneath the plate with thick, more buoyant continental crust.

Near their boundary, the plates can lock together for centuries, then suddenly let go, making giant earthquakes. If the seafloor rises or falls during an earthquake, giant sea waves ("tsunamis") are generated.

A wedge-shaped region ("accretionary wedge") forms between the converging plates as material is scraped off the top of the subducting plate.

Farther inland, the top of the subducting plate reaches depths where it "sweats" hot water. The rising water melts rock in its path, forming a curved line of volcanoes ("volcanic arc") on the overriding plate.

The region between the two mountain ranges ("forearc basin") remains near sea level.

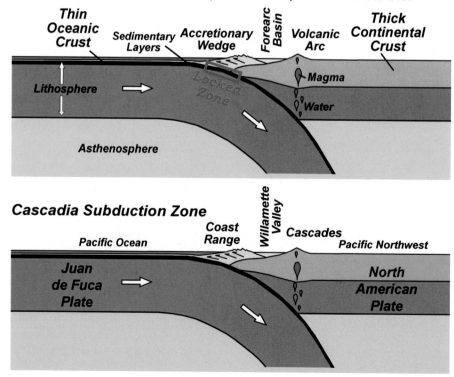

The Coast Range and Cascades are the two parallel mountain ranges in the Pacific Northwest. The intervening low area is the Willamette Valley in Oregon and Puget Sound in Washington.

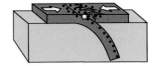

The Cascadia Subduction Zone is the Region where the Juan De Fuca and North American Plates Converge

The Coast Ranges, including the Olympic Mountains, are made of oceanic sediments and hard rocks that were caught in the vise between the converging plates, uplifted, and added to the edge of the continent.

Puget Sound and the Willamette Valley are areas near sea level between the coastal and volcanic mountain ranges.

The Cascades form above the line where the subducting plate extends to depths where it heats up, dehydrates and causes magma to form. The line of active volcanoes, from Mt. Garibaldi to Lassen Peak, coincides with the north-to-south extent of the subducting plate boundary.

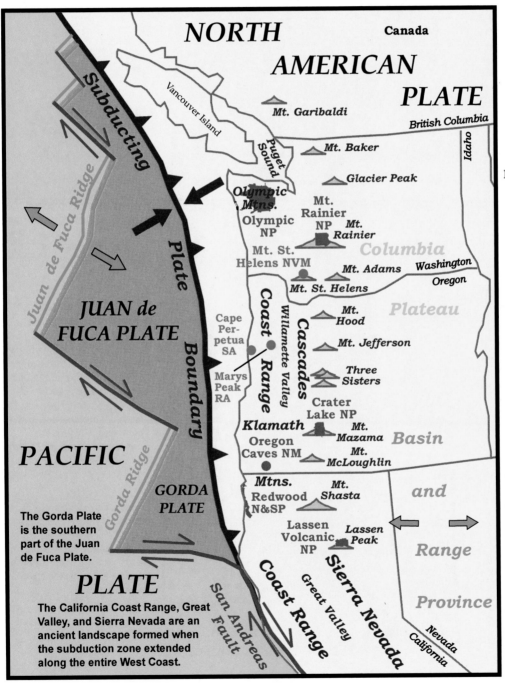

National Park Service

U. S. Forest Service

National Park Service and U. S. Forest Service sites in the region reveal the Beauty from the Beast of the Cascadia Subduction Zone. Their inspiring landscapes are products of the same tectonic forces responible for the earthquakes, tsunamis, volcanic eruptions and landslides that affect the region.

The Gorda Plate is the southern part of the Juan de Fuca Plate.

The California Coast Range, Great Valley, and Sierra Nevada are an ancient landscape formed when the subduction zone extended along the entire West Coast.

Plate Boundaries:

 Divergent

 Convergent

 *Transform*

Hotspot

Cascadia Subduction Zone Topography is Dominated by Two Parallel Mountain Ranges Separated by a Low Region Near Sea Level

The **COASTAL MOUNTAIN RANGES**, including the Olympic Mountains in northwest Washington and the Coast Range in southwest Washington, western Oregon and northern California, forms as sedimentary and volcanic layers are scraped off the top of the subducting oceanic plate and added to the edge of the continent.

The low region between the two parallel mountain ranges is the **PUGET SOUND** area in Washington, the **WILLAMETTE VALLEY** in Oregon, and the **GREAT VALLEY** in northern California.

150 miles inland, the top of the subducting plate reaches depths where it's hot enough to generate fluids, forming volcanoes in the **CASCADES**.

The **KLAMATH MOUNTAINS** are a block of thick crustal material ("accreted terrane") that crashed into the continent. This disrupted the subduction zone pattern (coastal range/valley/volcanoes) in southern Oregon and northern California.

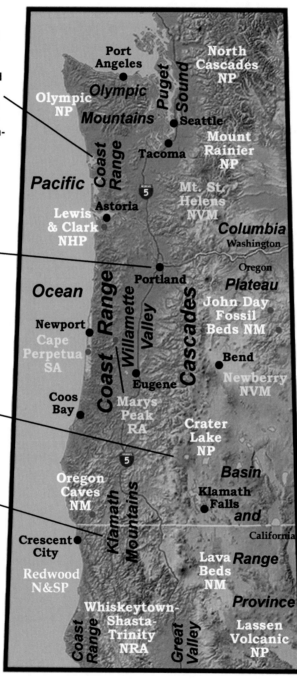

National Park Service

U. S. Forest Service

John Day Fossil Beds National Monument has volcanic material derived from ancient Cascade Volcanoes and the Columbia Plateau/Yellowstone Hotspot.

Although undoubtedly influenced by volcanic processes related to the Cascadia Subduction Zone, Newberry National Volcanic Monument, Lava Beds National Monument, and the eastern portion of Lassen Volcanic National Park also have landscapes and volcanic activity due to continental rifting in the Basin and Range Province.

Oregon Caves National Monument and North Cascades National Park have older rocks that are parts of terranes accreted to the North American continent during the past 200 million years.

The Topography of the Cascadia Subduction Zone Plays a Key Role in the Climate, Ecology and Human History of the Pacific Northwest

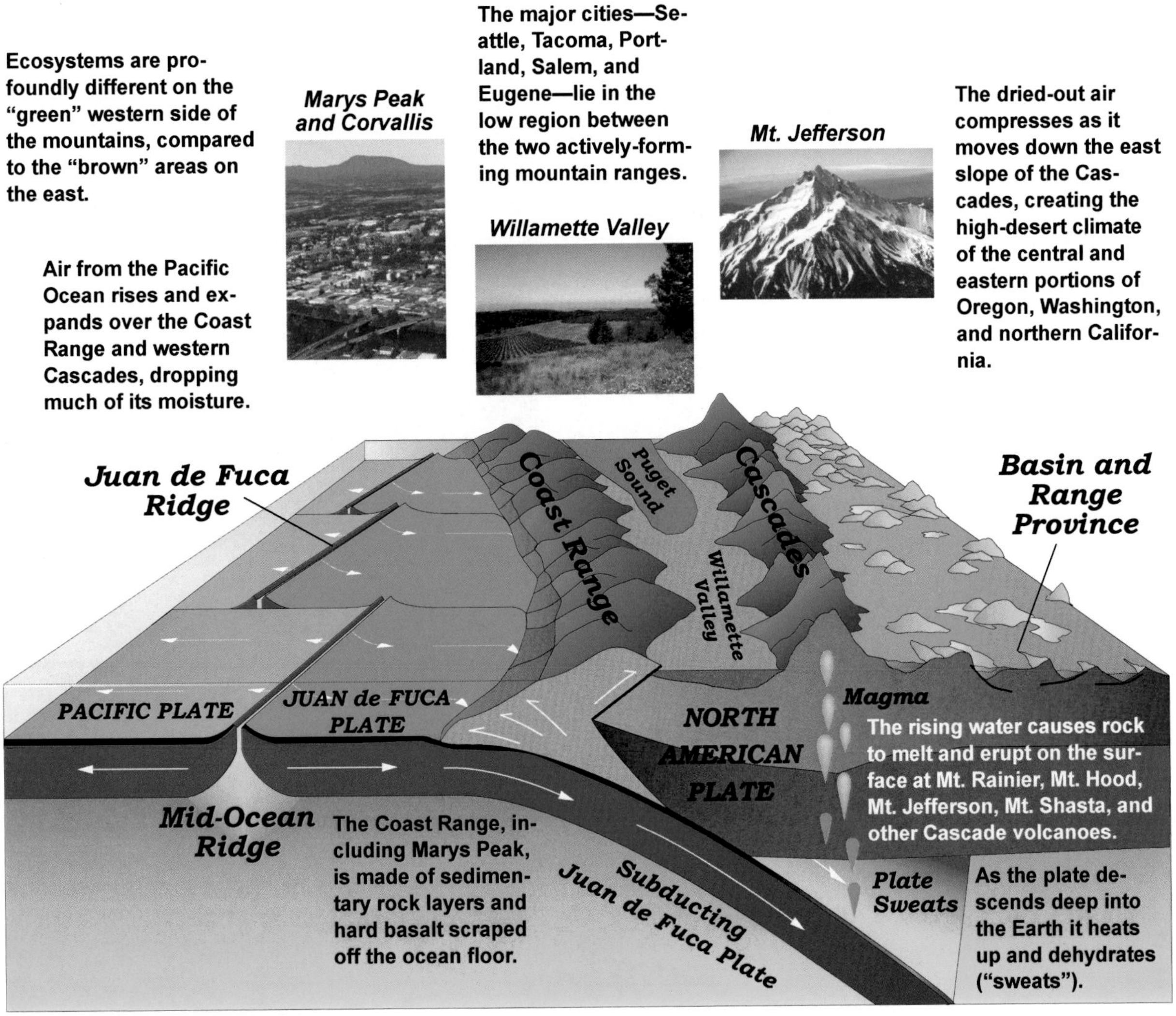

Ecosystems are profoundly different on the "green" western side of the mountains, compared to the "brown" areas on the east.

Air from the Pacific Ocean rises and expands over the Coast Range and western Cascades, dropping much of its moisture.

Marys Peak and Corvallis

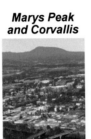

The major cities—Seattle, Tacoma, Portland, Salem, and Eugene—lie in the low region between the two actively-forming mountain ranges.

Willamette Valley

Mt. Jefferson

The dried-out air compresses as it moves down the east slope of the Cascades, creating the high-desert climate of the central and eastern portions of Oregon, Washington, and northern California.

Juan de Fuca Ridge

Coast Range

Puget Sound

Cascades

Willamette Valley

Basin and Range Province

Mid-Ocean Ridge

PACIFIC PLATE

JUAN de FUCA PLATE

NORTH AMERICAN PLATE

Magma

The rising water causes rock to melt and erupt on the surface at Mt. Rainier, Mt. Hood, Mt. Jefferson, Mt. Shasta, and other Cascade volcanoes.

The Coast Range, including Marys Peak, is made of sedimentary rock layers and hard basalt scraped off the ocean floor.

Subducting Juan de Fuca Plate

Plate Sweats

As the plate descends deep into the Earth it heats up and dehydrates ("sweats").

33

practical and spiritual aspects of Native Americans' lives. Later settlers coming on the Oregon Trail continued past central Oregon and Washington to the lush farmland of the Willamette Valley and Puget Sound, the low-lying region between the two rising mountain ranges.

Coast and Coastal Ranges

At 4,097 feet, Marys Peak, near Corvallis, is the highest point in the Oregon Coast Range. A drive, hike, or bicycle ride up Marys Peak is an opportunity to observe the landscape of one of the most active and spectacular landscapes anywhere on the face of the Earth—the Cascadia Subduction Zone. Standing on the top, you can see it all. The Cascade Mountains to the east lie above the position where the subducting Juan de Fuca Plate has reached about 50 miles depth. When the sky is especially clear you can see the line of Cascade volcanic peaks all the way from Mt. Rainier, 186 miles away in Washington, to Mt. McLoughlin, at a distance of 141 miles in southern Oregon. On the west, the green hills of the Coast Range extend to the blue waters of the Pacific Ocean. The horizon line, 50 to 100 miles offshore, roughly corresponds to where the Juan de Fuca Plate begins its descent beneath the edge of North America, lifting materials out of the sea to form the Coast Range.

Some geological processes act so slowly that we may think of the Earth as rock-solid and unmoving. But given enough time, the coast ranges lift up distances we can measure, while at the same time wind, rain, and the pounding of waves wear the landscape back down. Pillow lavas form on the ocean floor where erupting magma encounters cold sea water. When caught between the converging Juan de Fuca and North American plates, the pillows and other ocean layers can be lifted upward to more than 2,000 feet in elevation on Marys Peak and other coastal mountains. Layers of sand and mud, deposited on the ocean floor and later turned into the sedimentary rocks sandstone and shale, are seen above the pillows. Dramatic

Beauty of the Cascadia Coast

Olympic National Park, Washington

Robert J. Lillie

Oregon Islands National Wildlife Refuge, Oregon

Robert J. Lillie

Cape Blanco State Park, Oregon

Robert J. Lillie

Redwood National and State Parks, California

National Park Service

Formation of the Coast Range

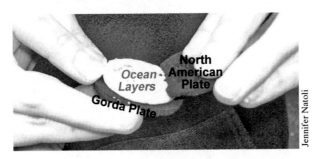

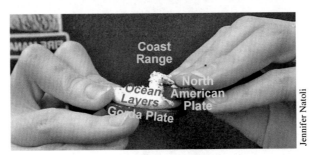

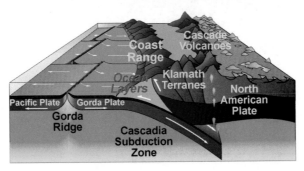

Ranger Jen's Oreo Demo. Jennifer Natoli was a seasonal ranger at Redwood National and State Parks. In her version of the Oreo® cookie demonstration, the creamy filling represents the layers of sediment and basalt on the ocean floor. As the Gorda Plate (lower cookie) subducts beneath the North American Plate (upper cookie), the layers are scraped off the ocean floor and pile up as the northern California Coast Range. A similar thing is happening farther north, where ocean layers scraped off the top of the Juan de Fuca Plate form the Oregon and Washington Coast Range and the Olympic Mountains.

Coast Range Layers Deposited on the Ocean Floor were Later Lifted Up and Plastered to the Edge of the Continent

**Marys Peak Recreation Area,
Siuslaw National Forest, Oregon**

Robert J. Lillie

**Olympic National Park,
Washington**

Robert J. Lillie

Tilted layers of thick sandstone (pink) and thin shale (dark) reveal the enormous forces that lifted and deformed the oceanic materials as the Juan de Fuca and North American plates converged.

Robert J. Lillie

Blobs of pillow basalt, some the size of watermelons, formed where hot lava erupted from the ocean floor (bottom). Layers of sand and mud—now sandstone and shale—were deposited over the basalt as the tectonic plate moved toward North America (top).

examples of these layers in tilted and contorted forms can be found in Olympic National Park and elsewhere along the coast and in the mountains. This contortion attests to the great forces that squeezed and lifted the layers from the sea.

Cascades

Cascade volcanoes lie above the region where the top of the subducting Juan de Fuca Plate reaches about 50 miles depth. Increased temperatures and pressures at that depth cause

Beauty of the Cascade Volcanoes

Mt. Rainier — Mt. Rainier National Park, Washington

Robert J. Lillie

Mt. Adams — Gifford Pinchot National Forest, Washington

Robert J. Lillie

Mt. Jefferson — Willamette, Deschutes, and Mt. Hood National Forests, Oregon

Robert J. Lillie

Three Sisters — Deschutes and Willamette National Forests, Oregon

Robert J. Lillie

37

Cascade Volcanoes: Active or Inactive?

Mt. Shasta, California

Robert J. Lillie

Active Volcano

A Cascade volcano that has been active since the last ice age has a cone shape with smooth slopes—lava flows and other volcanic materials smooth out the rough topography caused by glacial erosion.

Mt. Thielsen, Oregon

Robert J. Lillie

Inactive Volcano

A Cascade volcano that has not erupted since the last ice age has been deeply eroded by glacial ice and has a rough, jagged appearance—there are no recent volcanic deposits to smooth out the topography.

the rocks to metamorphose and dehydrate ("sweat"). The rising hot water causes overlying rock to melt, generating magma that at times erupts out on the surface.

The snow-capped High Cascades, including Mt. Rainier in Washington, Mt. Hood in Oregon, and Mt. Shasta in California, are juvenile geologic features. Most are younger than 700,000 years. Some High Cascade peaks erupted during the last ice age. Those that were quiet were worn down by glacial ice and form jagged peaks like Oregon's Mt. Washington and Mt. Thielsen. Others, like Mt. Rainier, Mt. Hood and Mt. Shasta, have rebuilt smooth cones as

volcanic material replaced what glaciers have worn away.

Oregon's crown jewel, Crater Lake, lies within the throat of a large volcano, Mt. Mazama, which erupted violently and collapsed in on itself 7,700 years ago. Oral tradition carried down by generations of Klamath People suggests that they may have witnessed this event. The state of Oregon quarter, adopted in 2005, not only depicts the beauty of this 7th deepest of the world's lakes, it also models its dimensions accurately—about 5 miles across (quarter's face) and nearly 2,000 feet deep (quarters thickness).

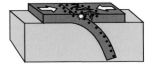

Crater Lake Partially Fills the Caldera of a Volcano that Erupted and Collapsed in on Itself 7,700 Years Ago

Mt. Mazama

Mt. Mazama was a composite volcano 10,000 to 12,000 feet in elevation. Its magma chamber was filled with heavy (low-silica) magma in its lower part and lighter (high-silica) magma above.

Paul Rockwood painting, courtesy National Park Service

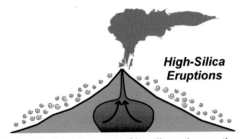

High-Silica Eruptions

During the initial stages of its climactic eruption, the upper part of the magma chamber poured out ash, pumice, and rhyolite lava flows.

Paul Rockwood painting, courtesy National Park Service

Collapse Caldera

So much magma erupted that the upper part of the mountain collapsed into the void, forming the caldera.

Paul Rockwood painting, courtesy National Park Service

Low-Silica Eruptions

Later eruptions of low-silica magma partially filled the caldera with basalt lava flows and cinder cones, including what is now Wizard Island. (The photo on the right is actually Aniakchak Volcano in Aniakchak National Monument and Preserve, a simiar collapsed volcano in Alaska).

National Park Service

Crater Lake

The coating of volcanic materials sealed the bottom of the caldera. Over a few centuries rain and snowmelt partially filled the hole with the waters of Crater Lake. At 1,943 feet, Crater Lake is the deepest lake in the United States and the 7th deepest in the world.

Robert J. Lillie

Living with the Beast

In the Pacific Northwest, as anywhere else, people learn to take the good with the bad. Part of the reason people live in and visit the Northwest is because of the beautiful and bountiful landscapes that surround them. The Cascadia Subduction Zone is largely responsible for the beauty. But the same plate convergence also results in the beast of earthquakes and volcanic eruptions, with their associated landslides, mudflows, and tsunamis. Panoramic views of the Cascadia Subduction Zone from Marys Peak and other special places help us envision these natural hazards and contemplate how we can better prepare our homes and communities.

Earthquakes and Tsunamis

Concerned residents and visitors have been pondering the question "When and where will the next big earthquake occur in the Pacific Northwest?" Research over the past four decades has revealed that giant earthquakes have occurred periodically, and that others will surely happen in the future. The last great earthquake occurred just over 300 years ago,

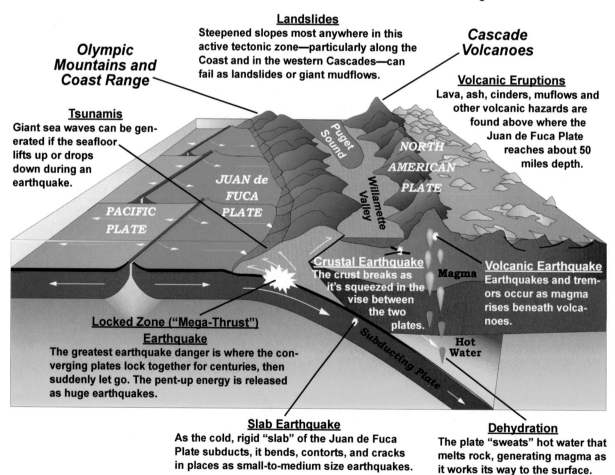

The "Beast" of the Cascadia Subduction Zone: Earthquakes, Tsunamis, Landslides and Volcanic Eruptions

Landslides
Steepened slopes most anywhere in this active tectonic zone—particularly along the Coast and in the western Cascades—can fail as landslides or giant mudflows.

Olympic Mountains and Coast Range

Cascade Volcanoes

Volcanic Eruptions
Lava, ash, cinders, muflows and other volcanic hazards are found above where the Juan de Fuca Plate reaches about 50 miles depth.

Tsunamis
Giant sea waves can be generated if the seafloor lifts up or drops down during an earthquake.

JUAN de FUCA PLATE

PACIFIC PLATE

NORTH AMERICAN PLATE

Crustal Earthquake
The crust breaks as it's squeezed in the vise between the two plates.

Magma

Volcanic Earthquake
Earthquakes and tremors occur as magma rises beneath volcanoes.

Locked Zone ("Mega-Thrust") Earthquake
The greatest earthquake danger is where the converging plates lock together for centuries, then suddenly let go. The pent-up energy is released as huge earthquakes.

Subducting Plate

Hot Water

Slab Earthquake
As the cold, rigid "slab" of the Juan de Fuca Plate subducts, it bends, contorts, and cracks in places as small-to-medium size earthquakes.

Dehydration
The plate "sweats" hot water that melts rock, generating magma as it works its way to the surface.

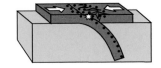

Solving the Earthquake Mystery

in the year 1700. Research reveals that the two converging tectonic plates have been locked together since then. When they suddenly unlock, the resulting earthquake could be devastating.

The zone where the plates are locked extends from offshore to beneath the coastal ranges. Most Pacific Northwest residents live in the Willamette Valley and Puget Sound regions, just inland from this zone. Hard, durable rock layers of the coastal ranges contrast with the soft sedimentary layers of the Willamette Valley and Puget Sound regions, which could shake like a bowl full of jelly during a large earthquake. That, coupled with landslides and tsunamis (large sea waves) generated by the earthquake, could devastate coastal regions and cut them off from food, medicine, and other supplies found in the more populated regions of the inland valleys.

The Cascadia Subduction Zone is a three-dimensional region that includes much of the Pacific Northwest. A variety of earthquakes shake this region due to plate convergence. The largest occur where the Juan de Fuca and North American plates are locked together, as they have been for the past three centuries. When the plates suddenly let go, a massive earthquake will shake the entire Pacific Northwest, a series of tsunami waves will pound the Coast, and landslides will make it difficult to reach some of those in need. But the moving plates can cause destruction in other ways. Intra-plate (or "slab") earthquakes, like the magnitude 6.8 Nisqually Earthquake that struck Washington's Puget Sound region in 2001, occur within the subducting Juan de Fuca Plate. Populations in the Portland and Seattle areas are also at risk from crustal earthquakes at shallow depths on the North American Plate, such as the 1993 magnitude 5.6 Scotts Mills Earthquake beneath the western Willamette Valley in Oregon. And as magma moves to shallow depths beneath volcanoes, small earthquakes and tremors can shake the Cascades.

The most recent great earthquake on the Cascadia Subduction Zone occurred in the year 1700 AD. There were no written records kept in the Pacific Northwest at that time to document the event. But geologists are detectives. Using not only scientific observations and analyses, but also historical and anthropological evidence, they were able to piece together clues that unraveled the mystery. These clues suggest that there is about a one-in-three chance that the next Great Cascadia Subduction Zone earthquake—and accompanying tsunami—will strike the Pacific Northwest in the next 50 years.

Prior to the 1970's, it wasn't thought that the Pacific Northwest was in danger of earthquakes with magnitudes much larger than the low 7's. But the move to build nuclear reactors in the region led to a debate among scientists about a much larger earthquake risk from the Cascadia Subduction Zone. Seismologists noted that Cascadia was much like other subduction zones, except for very few earthquakes. Maybe the reason for the small number of earthquakes is that the subduction zone is completely locked, but ready to let go as a giant earthquake?

Several lines of evidence emerged that show that the Cascadia subduction zone has large earthquakes and accompanying tsunamis about every 200 to 600 years, and that the last one occurred on January 26, 1700. There are standing trees along the Oregon and Washington coasts that are thought to have died when they were inundated by sea water as areas dropped down during the earthquake. Carbon dating of the wood reveals that the trees died about 300 years ago, although resolution of the date is only about ± 20 years. Closer examination by identifying and counting tree rings shows that the trees died between the fall of 1699 and the spring of 1700.

Historical evidence from Japan offers an even more precise date for the earthquake. In Japan, giant sea waves, known as tsunamis, commonly occur just after a large earthquake

Convergent Plate Boundary

Cascadia Subduction Zone:
Geological Evidence for Past and Future Earthquakes and Tsunamis

Niawiakum River, Washington

The author points to the sand layer deposited by a tsunami spawned by the last Great Cascadia Earthquake in the year 1700.

U. S. Geological Survey geologist Brian Atwater (yellow coat) engages coastal educators about Cascadia earthquake and tsunami geology. A core drilled into the nearby tidal marsh reveals the 1700 tsunami sand.

Elk River Estuary, Washington

Tree ring expert David Yamaguchi stands in front of a tree killed by salt water when the land surface dropped during the 1700 earthquake. The slice from a nearby western red cedar that survived shows normal growth rings up to the year 1699, then stunted growth caused by the salt-water invasion.

Robert J. Lillie

Robert J. Lillie   Robert J. Lillie   Robert J. Lillie

Robert J. Lillie   Robert J. Lillie

Robert J. Lillie (Niawiakum photos)

shakes the region. But occasionally an "orphan" tsunami arrives that was caused by a distant earthquake. Precise records from such a tsunami that struck Japan have allowed scientists to conclude that the source was a large Pacific Northwest earthquake occurring on January 26, 1700. Many shallow bays and estuaries along the Oregon and Washington coast also contain sand layers thought to have been deposited by the 1700 tsunami.

In many areas along the Washington, Oregon, and northern California coasts one can find evidence of marsh plants, including trees, that were suddenly killed and partially buried by deposits of sand. Above the sand are bay muds and then more marsh and sand deposits. This sequence repeats at least seven times over the past 3,000 years. Each sequence may indicate the time between large earthquakes. As the plates lock and the surface rises, shallow bay areas rise up and become marshes. Then, when the earthquake occurs as the plates suddenly unlock, the surface abruptly drops below sea level and is covered by sand from the resulting tsunami waves. Over the next few hundred years the area slowly rises up again until the next big earthquake.

Recent studies involving the placement of GPS instruments also provide compelling evidence that the Cascadia Subduction Zone is "locked and loaded." We are used to GPS instruments in our cars that tell us that we are at a particular intersection. Such instruments have a resolution of about 10 feet. But the scientific GPS instruments installed in Cascadia have a resolution of about one millimeter (4/100 of an inch—the width of a pinhead!). And instead of moving from one point on Earth's surface to another, the instruments are anchored firmly into hard rock. They thus show the movement

The Newport, Oregon GPS Station Shows that the Coastal Region is being Pushed Northeastward by the Juan de Fuca Plate

Robert J. Lillie

Educators Roger Groom (left) and Bob Butler point northeastward. The GPS measurements reveal that the Newport region has been moving in that direction since the instrument was installed more than a decade ago.

The dots on the graph show that the Newport region has moved northeastward 5 inches in 13 years—an average of 0.4 inches per year. At that rate, the region has moved more than 10 feet since the 1700 earthquake.

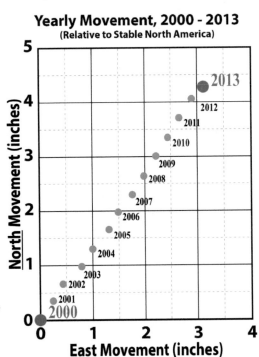

Development of a Cascadia Mega-Thust Earthquake

Region After Previous Large Earthquake

The Juan de Fuca Plate subducts in a
northeast direction beneath the North
American Plate.

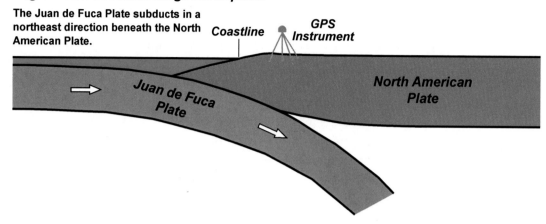

Plates Lock Together for Centuries

The coastal region
is pushed upward
and northeastward.

Forests can grow
in areas lifted out
of the sea.

GPS instrument registers the
upward and northeastward motion.

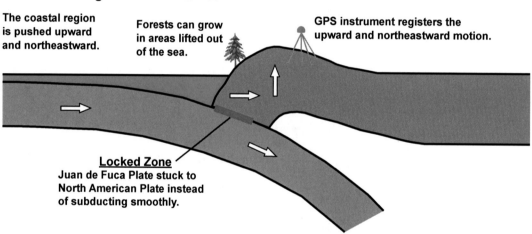

Locked Zone
Juan de Fuca Plate stuck to
North American Plate instead
of subducting smoothly.

Plates Suddenly Unlock

A magnitude 8 to 9
earthquake shakes
the Pacific Northwest.

Trees in down-
dropped areas killed
by seawater.

GPS instrument drops down and measures
the sudden southwestward motion of the
coastal region, back toward the ocean.

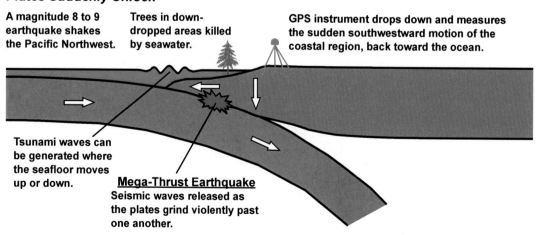

Tsunami waves can
be generated where
the seafloor moves
up or down.

Mega-Thrust Earthquake
Seismic waves released as
the plates grind violently past
one another.

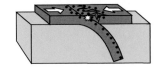

of Earth's surface itself.

GPS instruments along the Cascadia coast show that the region is moving about 1/2 inch per year toward eastern Oregon and Washington. If the Juan de Fuca Plate were subducting smoothly beneath the edge of North America, the coast would not be moving in such a fashion. But because the edge of North America is stuck to the Juan de Fuca Plate, it is being squeezed upward and northeastward. If the plates have been stuck since the last earthquake over 300 years ago, at that rate they would have moved over 10 feet toward the northeast. When the plates suddenly let go, the edge of North America might suddenly lunge 10 feet seaward, grinding violently as a magnitude 8 to 9 earthquake. And if the seafloor moves up or down in the process, a tsunami will be generated. We could expect 2 to 5 minutes of violent shaking followed by giant sea waves less than half an hour later.

Coupled with Native American stories about large earthquakes and tsunamis that occurred several generations ago, the evidence is overwhelming that massive earthquakes are a common occurrence in the Pacific Northwest and that we have entered the window of opportunity for one to occur in the not-too-distant future.

Tsunamis in the Pacific Northwest

The massive earthquakes and tsunamis that devastated the Indian Ocean region in 2004 and Japan in 2011 have heightened awareness that the Northwest Coast is prone to tsunamis from local and distant sources. A tsunami produced by an earthquake far from the Northwest takes several hours to cross the ocean, allowing time for an official warning and evacuation. But a tsunami from an earthquake on the nearby Cascadia Subduction Zone will reach the coast in 20 to 30 minutes. People on open beaches, in low-lying areas, tidal flats, or near mouths of rivers draining into the ocean are in the greatest danger from a tsunami. If you're on the Coast and feel strong shaking from an earthquake, or see the water suddenly start to come in or move out, get to higher ground immediately!

The horrific earthquake and tsunami that struck Japan on March 11, 2011 occurred in a region similar to the Cascadia Subduction Zone.

Native American Stories of Earthquakes and Tsunamis

Although Native Americans don't have a written record of their history, their oral tradition is rich with vivid accounts of events that have shaped the lives of their ancestors. The following story was told to Jason T. Younker by his uncle. Jason published it in the Summer 2007 issue of the Oregon Historical Quarterly.

<u>Weaving Long Ropes</u>

It wasn't too long ago when our people learned a great lesson, and a great tide would sweep many of them away. They were told by their elders "to weave long ropes because you never know when a big tide is coming and you won't have much time. If you don't have long ropes when the waters rise you'll be swept away." They were warned over and over, but few paid attention. Soon after, an offshore earthquake made a big tide. The waters rushed up the valleys and quickly overwhelmed many in the villages. Many of the people were unable to make it to their canoes. The water rose higher and higher until the tops of the tallest trees were visible. Those who had ropes quickly tied their canoes to the treetops. Soon all the trees were covered by the rising water. ... Only a few remained, and these were the wiser ones who had heeded the elders' warnings.

Off the east coast of Japan, the Pacific Plate is diving westward beneath the edge of the Asian continent. When the plates suddenly let go after being stuck for about 1,000 years, a magnitude 9.0 earthquake struck the region. Japan is a prosperous nation with infrastructure that is as prepared as any country to withstand the enormous and prolonged shaking. And indeed, for the most part, buildings, bridges, and other structures faired well during the five minutes of extreme earthquake shaking. But unfortunately, as the Asian Plate lunged 8 feet eastward over the Pacific Plate, it uplifted the ocean floor, sending a tsunami that struck northeast Japan with horrific results, killing more than 15,000 people and displacing millions. It also sent a "wake-up call" to the Pacific Northwest, as waves raced across the ocean and damaged the Oregon and California coasts.

Destructive distant tsunamis have struck the Pacific Northwest from as far away as Alaska and Chile. Since 1854, 21 tsunamis produced by earthquakes around the Pacific Ocean reached the Northwest coast. The Pacific Northwest was affected by a tsunami caused by the 1964 Great Alaska Earthquake. Traveling the speed of a jetliner—about 500 miles per hour—even after traveling 3,000 miles, the waves lifted bridges from their moorings, ripped roofs from buildings, and did considerable damage to an elementary school building in Cannon Beach, Oregon—which, fortunately, was not in session. It took about four hours for the initial waves to travel from Alaska. The waves resulted in the deaths of 11 people in Crescent City, California, and four children camping with their family at Beverly Beach, Oregon.

Volcanoes

The Cascade volcanoes extend in a line from north-to-south, all the way from Mt. Garibaldi in southern British Columbia to Lassen Peak in northern California. These volcanoes, each with their own personalities, tend to act up at times like spoiled children or angry adults.

Part of the reason for the wide variety in their eruption styles and histories is the complexity of magmas generated in an ocean-continent subduction zone.

When a plate capped by thin oceanic crust (such as the Juan de Fuca Plate) descends downward, it is subjected to heat and pressure. This causes the rock to metamorphose—that is, its minerals recrystallize to other forms. A by-product of this recrystallization is the release of fluids, particularly water. In other words, just like us, when the plate is under a lot of pressure and can't stand the heat, it sweats! Some of that water rises up into the overriding plate (in this case, the North American Plate). When the rising water encounters the already hot mantle of that plate, it causes minerals in the rock to melt. The magma that is produced in this process has relatively low-silica, basalt composition.

As the basalt magma rises up through the thick continental crust of North America, it melts some of that rock, too. In doing so it becomes enriched in silica, as high-silica minerals commonly are the first ones to melt. The resulting magmas can therefore have a variety of compositions, ranging from the original (low-silica) basalt, to (intermediate-silica) andesite, all the way up to (high-silica) rhyolite. Magmas that ultimately make it to the shallow crust can settle into lower pockets with heavy, low-silica basalt magma, overlain by higher-silica andesite and rhyolite. Solids and gases are also part of the mix.

Magmas high in silica are thick and pasty. Gases may be trapped in them under high pressure. Thus subduction-zone volcanoes—including those in the Cascades—can be explosive (think of the release of steam in a boiling pot of thick oatmeal). Their high-silica magmas produce not only andesite and rhyolite lava flows, but also lots of other volcanic materials such as ash and pumice. When mixed with water from glaciers, snowmelt, and streams, these materials can make very dense, fast-moving volcanic mudflows (known sometimes by the Indone-

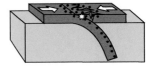

Cascade Volcanoes Have Their Own Beauty and Personalities

Mt Garabaldi

Creative Commons

**Last Erupted:
6,000 Years Ago**

Mt. Baker

USGS

**Last Erupted:
6,700 Years Ago**

Glacier Peak

USGS

**Last Erupted:
1,100 Years Ago**

Mt. Rainier

USGS

**Last Erupted:
1,000 Years Ago**

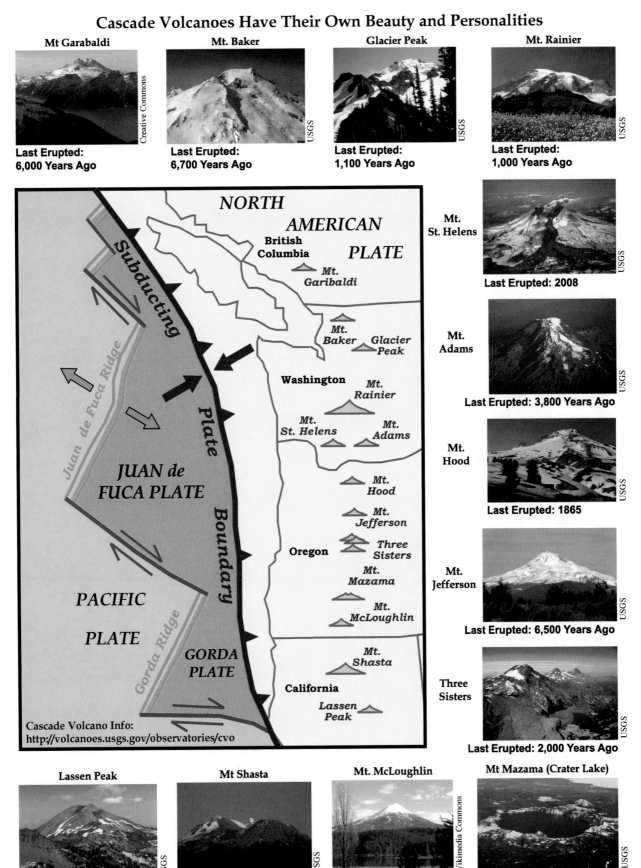

**Mt.
St. Helens**

USGS

Last Erupted: 2008

**Mt.
Adams**

USGS

Last Erupted: 3,800 Years Ago

**Mt.
Hood**

USGS

Last Erupted: 1865

**Mt.
Jefferson**

USGS

Last Erupted: 6,500 Years Ago

**Three
Sisters**

USGS

Last Erupted: 2,000 Years Ago

Lassen Peak

USGS

Last Erupted: 1917

Mt Shasta

USGS

Last Erupted: 1786

Mt. McLoughlin

Wikimedia Commons

**Last Erupted:
30,000 Years Ago**

Mt Mazama (Crater Lake)

USGS

**Last Erupted:
6,600 Years Ago**

47

Eruption of Mt. St. Helens, Washington

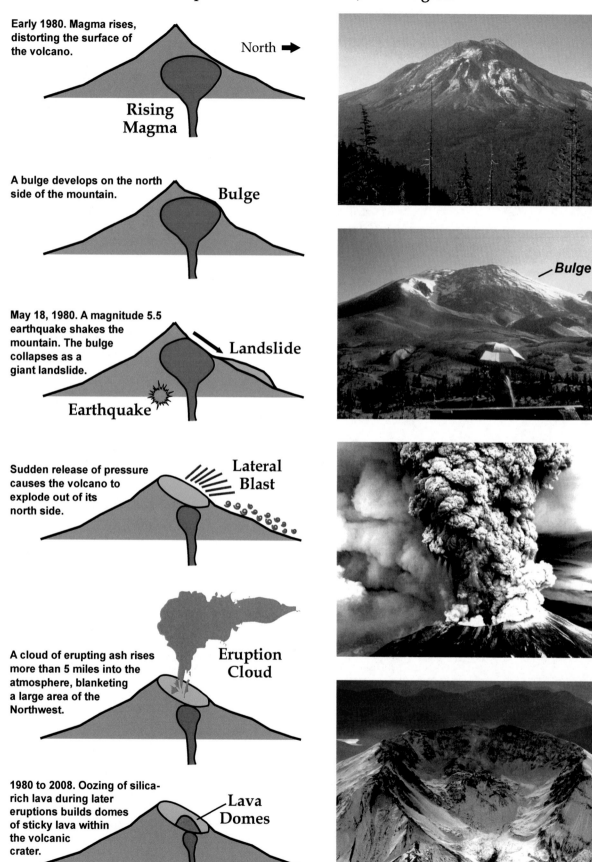

Early 1980. Magma rises, distorting the surface of the volcano.

North ➡

Rising Magma

A bulge develops on the north side of the mountain.

Bulge

May 18, 1980. A magnitude 5.5 earthquake shakes the mountain. The bulge collapses as a giant landslide.

Landslide

Earthquake

Sudden release of pressure causes the volcano to explode out of its north side.

Lateral Blast

A cloud of erupting ash rises more than 5 miles into the atmosphere, blanketing a large area of the Northwest.

Eruption Cloud

1980 to 2008. Oozing of silica-rich lava during later eruptions builds domes of sticky lava within the volcanic crater.

Lava Domes

Bulge

U. S. Geological Survey

U. S. Geological Survey

U. S. Geological Survey

U. S. Geological Survey

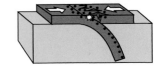

sian term "lahar"). And at times, lower-silica components of their magma chambers erupt, producing basalt lava flows as well as cinders and volcanic bombs.

Many Pacific Northwest residents were around when Mt. St. Helens erupted on May 18, 1980. They witnessed not only the effects of its ash over eastern Washington, Idaho, and Montana, but also slurries of mud that roared down small streams in Washington on their way to the Columbia River. Much larger volcanic mudflows are a threat to communities on the flanks of Mt. Rainier and Mt. Hood.

Rapidly growing populations are infringing on volcanic areas, increasing the potential for disaster. Volcanoes in the Cascade mountain range threaten communities in the Pacific Northwest. Several volcanoes have erupted recently, or show signs that they will erupt in the not-too-distant future. Scientific studies of the Cascades help us appreciate volcanic hazards and how they affect our lives. Most importantly, this research helps us plan our communities and prepare our homes and families for inevitable eruptions and their after effects.

Additional Reading

Alt, David D. and Donald W. Hyndman, 1978, "Roadside Geology of Oregon," Missoula, Mt: Mountain Press Publishing Company, 278 pp.

Alt, David D. and Donald W. Hyndman, 1984, "Roadside Geology of Washington," Missoula, MT: Mountain Press Publishing Company, 288 pp.

Alt, David D. and Donald W. Hyndman, 2000, "Roadside Geology of Northern and Central California," Missoula, MT: Mountain Press Publishing Company, 384 pp.

Atwater, Brian F., 1992, "Geologic evidence for earthquakes during the past 2000 years along the Copalis River, southern coastal Washington," Journal of Geophysical Research, v. 97, p. 1901-1919.

Atwater, Brian, F., 2005, "The Orphan Tsunami of 1700: Japanese Clues to a Parent Earthquake in North America," Seattle, WA: University of Washington Press, 144 pp.

Bishop, Ellen M., 2003, "In Search of Ancient Oregon," Portland, OR: Timber Press, 288 pp.

Bishop, Ellen M., 2014, "Living with Thunder: Exploring the Geologic Past, Present, and Future of the Pacific Northwest," Corvallis, OR: OSU Press, 160 pp.

Clark, E. E., 1953, "Indian Legends of the Pacific Northwest," University of California Press, Berkeley, California, 225 p.

Decker, R., and B. Decker, 2001, "Volcanoes in America's National Parks," New York: W. W. Norton and Comp., 256 pp.

Doughton, Sandi, 2014, "Full Rip 9.0: The next big earthquake in the Pacific Northwest," Seattle, WA, Sasquatch Books, 288 pp.

Harris, Stephen L., 2005, "Fire Mountains of the West: The Cascade and Mono Lake Volcanoes," Missoula, MT: Mountain Press Publishing Company, 3rd Edition, 453 pp.

Henderson, Bonnie, 2014, "The next Tsunami: Living on a Restless Coast," Corvallis, OR: Oregon State University Press, 330 pp.

Hoblitt, R. P., Walder, J. S., Driedger, C. L., Scott, K. M., Pringle, P. T., and Vallance, J. W., 1998, "Volcano Hazards from Mount Rainier, Washington," Washington, DC: U. S. Geological Survey, Open-file Report 98-428, 11 p., 1 pl.

Loomis, B. F., 1926, "Pictorial History of the Lassen Volcano," Mineral, California: Loomis Museum Association, 96 pp.

Miller, Marli B., 2014, "Roadside Geology of Oregon," 2nd Ed, Missoula, Mt: Mountain Press Publishing Company, 380 pp.

Orr, Elizabeth L. and William N. Orr, 2000, "Oregon Geology," 6th Edition, Corvallis, OR: Oregon State University Press, 304 pp.

Rau, W. W., 1973, "Geology of the Washington Coast between Point Grenville and the Hoh River," Washington Department Natural Resources, Geology and Earth Resources Division, Bulletin 66, 58 pp.

Rau, W. W., 1980, "Washington Coastal Geology between the Hoh and Quillayute Rivers," Washington Department Natural Resources, Geology and Earth Resources Division, Bulletin 72, 57 pp.

Scott, K. M., and Vallance, J. K., 1995, "Debris flow,

debris avalanche, and flood hazards at and downstream from Mount Rainier," Washington, DC: U. S. Geological Survey, Hydrologic Investigation Atlas HA-729, 1:100,000, 9 pp., 2 pl.

Scott, K. M., and Vallance, J. W., and Pringle, P. T., 1995, "Sedimentology, behavior, and hazards of debris flows at Mount Rainier," Washington, DC: U. S. Geological Survey Professional Paper 1547, 56 pp.

Sisson, T. W., 1995, "History and hazards of Mount Rainier, Washington: Volcano hazards fact sheet," Washington, DC: U. S. Geological Survey Open-file Report 95-642, 2 pp.

Sisson, T. W., J. W. Vallance, and P. T. Pringle, 2001, "Progress made in understanding Mount Rainier's Hazard's," EOS: Transactions, American Geophysical Union, v. 82, p. 113-120.

Sullivan, William L., 2008, "Oregon's Greatest Natural Disasters," Eugene, OR: Navillus Press, 263 pp.

Tabor, R. W., 1987, "Geology of Olympic National Park," Seattle: Northwest Interpretive Association, 144 pp.

Tilling, R. I., 1982, "Eruptions of Mount St. Helens: Past, Present, and Future," Washington, DC: United States Government Printing Office, 46 pp.

Yeats, Robert S., 2004, "Living with Earthquakes in the Pacific Northwest," 2nd Ed., Corvallis, OR: Oregon State University Press, 390 pp.

Younker, Jason T., 2007, "Weaving Long Ropes: Oral Tradition and Understanding the Great Tide," Oregon Historical Quarterly, v. 108, no. 2, p. 193-200.

3

Transform Plate Boundary: San Andreas Fault

California's sheared-up landscape and earthquake hazards reflect the movement of the Pacific Plate past the edge of North America. A transform plate boundary, where one tectonic plate slides laterally past another, extends through western California from the Mexican border to the Pacific Northwest. This feature includes the famous San Andreas Fault, responsible not only for destructive earthquakes, but also for the spectacular scenery of the San Francisco Bay area and other coastal regions of California.

How Transform Motions Affect the Pacific Northwest

The Pacific Northwest is affected by transform plate boundary motion because of its proximity to the San Andreas Fault. Earthquakes

San Andreas Transform Plate Boundary

Golden Gate National Recreation Area, California

Robert J. Lillie

The Golden Gate Bridge spans the entrance to San Francisco Bay. In the background, the city of San Francisco lies on a peninsula dissected by the San Andreas Fault, along which the western portion of the city slides northward past the eastern part.

Pt. Reyes National Seashore, California

Robert J. Lillie

Participants in a geology interpretive workshop straddle the San Andreas Fault along the Earthquake Trail. Those in the foreground point northward to represent the motion of the Pacific Plate past the edge of North America. Participants in the background are on the North American Plate and point southward.

Transform Plate Boundaries Affect the Pacific Northwest and California — Both Onshore and Offshore

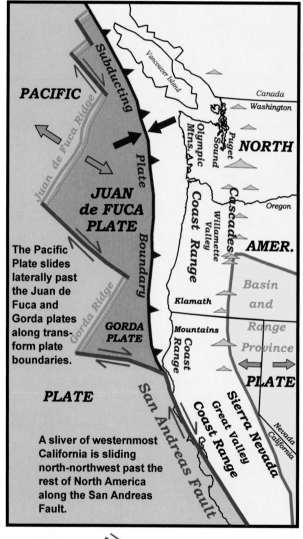

PACIFIC

Juan de Fuca Ridge

Subducting

Vancouver Island

Canada
Washington

Olympic Mtns.
Puget Sound

NORTH

JUAN de FUCA PLATE

Plate Boundary

Coast Range

Willamette Valley

Cascades

Oregon

AMER.

The Pacific Plate slides laterally past the Juan de Fuca and Gorda plates along transform plate boundaries.

Gorda Ridge

GORDA PLATE

Klamath

Mountains

Coast Range

Basin
and
Range
Province

PLATE

PLATE

San Andreas Fault

Sierra Nevada
Great Valley
Coast Range

Nevada
California

A sliver of westernmost California is sliding north-northwest past the rest of North America along the San Andreas Fault.

Plate Boundaries: *Transform* — *Divergent* — **Convergent**

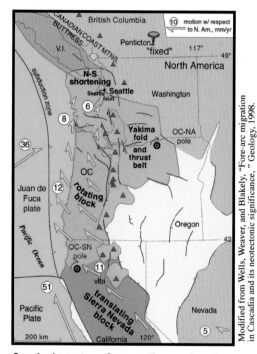

San Andreas transform motion pushes the Sierra Nevada crustal block northward, rotating blocks clockwise in the Coast Range, squeezing rocks in the Yakima fold and thrust belt, and influencing earthquake activity along fault lines such as the Seattle Fault.

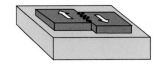

along this major tectonic feature can greatly upset cities along its length, including the San Diego, Los Angeles, and San Francisco/Oakland areas. But it's interesting to note that this region is so heavily populated because of the same tectonic forces that sometimes shake it up with such violent consequences during earthquakes. These forces also create a sheared-up landscape that includes spectacularly-beautiful coastlines and economically-important harbors. Thousands of earthquakes over millions of years have built this landscape not only along the major fault line—the San Andreas—but also on other faults within a broad zone of shearing between the Pacific and North American plates.

But the Northwest is also affected by transform plate motion in more direct ways. As the Pacific Plate slides northward past California, it pushes a block of rigid crust against the Klamath Mountains. This compression causes big blocks of the Coast Range of Oregon to move northward and rotate in a clockwise direction. The motion is partly responsible for small-to-moderate size earthquakes, as well as a zone of deformed rocks in southern Washington known as the Yakima fold and thrust belt.

There are also transform plate boundaries offshore from the Pacific Northwest. These zones of lateral motion connect mid-ocean ridge segments where the Pacific Plate diverges from the Juan de Fuca and Gorda plates. Pacific Northwest residents often wake up to radio or internet reports of earthquakes a few hundred miles off the coast. These earthquakes—which can be up to magnitude 6.5 or so—are not on the subducting plate boundary, but rather they occur along the transform plate boundaries, or where the tiny Gorda Plate is being squeezed by the other plates.

California Tectonics

About 200 million years ago, a large tectonic plate (called the Farallon Plate) started

West Coast Tectonic Evolution

40 Million Years Ago

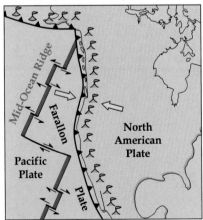

20 Million Years Ago

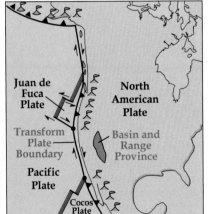

Today

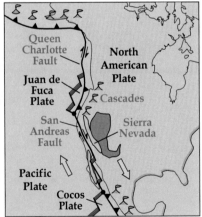

Forty million years ago, a large tectonic plate, known as the Farallon Plate, was between the Pacific and North American plates. Subduction of the Farallon Plate beneath the entire West Coast created a line of volcanoes from Alaska to Central America.

As the mid-ocean ridge separating the Farallon and Pacific Plates entered the subduction zone, the Farallon Plate separated into the Juan de Fuca and Cocos Plates. A transform plate boundary developed where the Pacific Plate was in contact with the North American Plate and the volcanism ceased in central California. Farther east, the continent began to rift apart in the Basin and Range Province.

The Cascades are the modern volcanic arc developing where the Juan de Fuca Plate subducts beneath the North American Plate. The Sierra Nevada are the eroded remnants of the volcanic arc developed when the Farallon Plate subducted beneath the continent. The San Andreas Fault and Queen Charlotte Fault are transform plate boundaries developing where the Pacific Plate moves northward past the North American Plate.

A Deck of Cards Demonstrates the Zone of Shearing at a Transform Plate Boundary

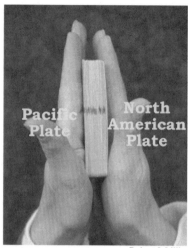

Robert J. Lillie

For western California, your left hand represents the rigid Pacific Plate, while your right hand is like the unaffected part of the North American Plate.

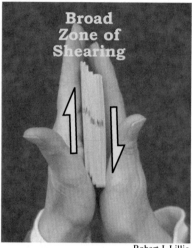

Robert J. Lillie

As you slide your hands laterally past one another, a broad zone of shearing develops as several card faces slip.

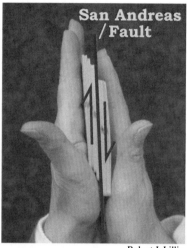

Robert J. Lillie

Eventually the weakest card face —the San Andreas Fault—dominates within the broad transform plate boundary.

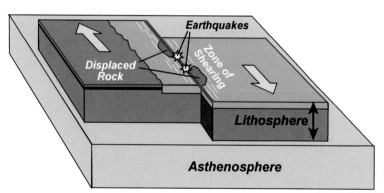

The broad zone of shearing at a transform plate boundary includes masses of rock displaced tens to hundreds of miles, shallow earthquakes, and a landscape consisting of long ridges separated by narrow valleys.

U. S. Geological Survey

The San Andreas Fault is just one of many active earthquake faults in a broad zone of shearing along the transform plate boundary in the San Francisco Bay Area.

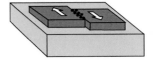

The San Andreas Fault is part of a Transform Plate Boundary that Disrupts the Topography of an Ancient Subduction Zone

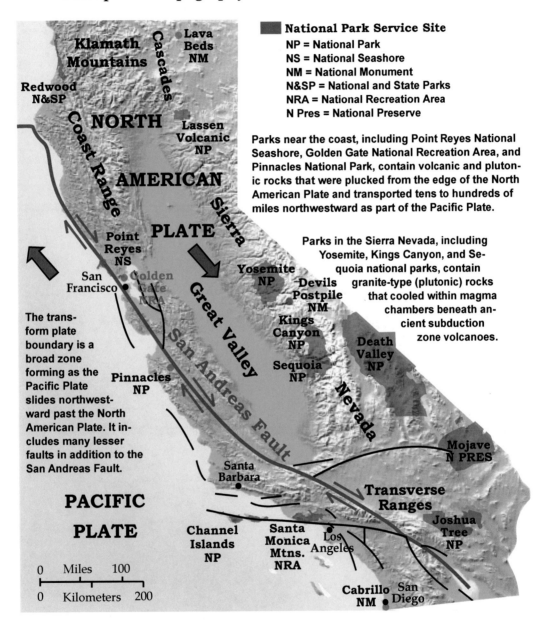

■ **National Park Service Site**
NP = National Park
NS = National Seashore
NM = National Monument
N&SP = National and State Parks
NRA = National Recreation Area
N Pres = National Preserve

Parks near the coast, including Point Reyes National Seashore, Golden Gate National Recreation Area, and Pinnacles National Park, contain volcanic and plutonic rocks that were plucked from the edge of the North American Plate and transported tens to hundreds of miles northwestward as part of the Pacific Plate.

Parks in the Sierra Nevada, including Yosemite, Kings Canyon, and Sequoia national parks, contain granite-type (plutonic) rocks that cooled within magma chambers beneath ancient subduction zone volcanoes.

The transform plate boundary is a broad zone forming as the Pacific Plate slides northwestward past the North American Plate. It includes many lesser faults in addition to the San Andreas Fault.

Subduction Zone Rocks in California Moved Long Distances Along the San Andreas Fault

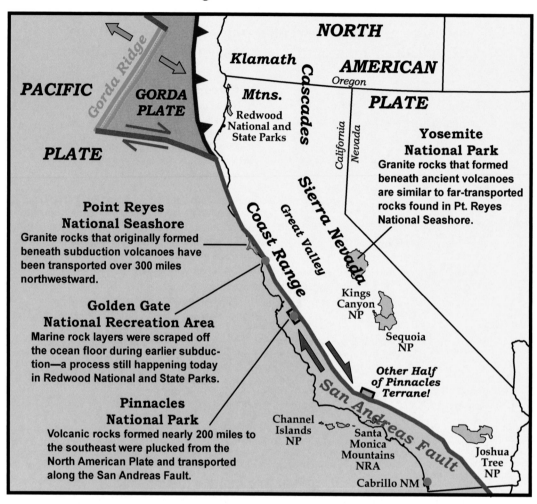

NORTH AMERICAN PLATE

Oregon

Klamath Mtns.

Cascades

California / Nevada

Sierra Nevada

Great Valley

Coast Range

PACIFIC PLATE

Gorda Ridge

GORDA PLATE

Redwood National and State Parks

Yosemite National Park
Granite rocks that formed beneath ancient volcanoes are similar to far-transported rocks found in Pt. Reyes National Seashore.

Point Reyes National Seashore
Granite rocks that originally formed beneath subduction volcanoes have been transported over 300 miles northwestward.

Golden Gate National Recreation Area
Marine rock layers were scraped off the ocean floor during earlier subduction—a process still happening today in Redwood National and State Parks.

Pinnacles National Park
Volcanic rocks formed nearly 200 miles to the southeast were plucked from the North American Plate and transported along the San Andreas Fault.

Kings Canyon NP

Sequoia NP

Other Half of Pinnacles Terrane!

San Andreas Fault

Channel Islands NP

Santa Monica Mountains NRA

Joshua Tree NP

Cabrillo NM

Ancient Subduction Zone

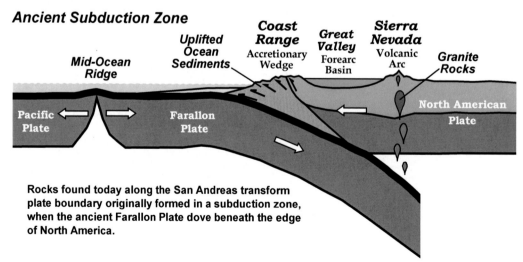

Coast Range Accretionary Wedge

Great Valley Forearc Basin

Sierra Nevada Volcanic Arc

Uplifted Ocean Sediments

Mid-Ocean Ridge

Granite Rocks

Pacific Plate

Farallon Plate

North American Plate

Rocks found today along the San Andreas transform plate boundary originally formed in a subduction zone, when the ancient Farallon Plate dove beneath the edge of North America.

56

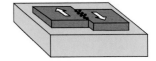

San Francisco Bay Area Parks Reveal a Sheared-Up, Ancient Subduction Zone Landscape

Pt. Reyes National Seashore, California

U. S. Geological Survey

Robert J. Lillie

The granite rocks are similar to those found in Yosemite National Park. They formed beneath ancient subduction zone volcanoes, were plucked from the edge of the North American Plate, and transpored more than 300 miles northwestward along the San Andreas Fault.

An offset fence line reveals the 16 feet of lateral ground breakage that occured as the San Andreas Fault suddenly let loose during the 1906 San Francisco Earthquake.

Golden Gate National Recreation Area, California

Long ridges and valleys around San Francisco Bay are in the zone of shearing along the transform plate boundary.

Robert J. Lillie

Robert J. Lillie

Robert J. Lillie

Pillow basalt, formed as lava poured out on the ocean floor, was later scraped off the top of the subducting plate and thrust onto the edge of the continent.

Layers of ocean sediment were squeezed and contorted as they were caught in the vise of the converging plates at the ancient subduction zone.

57

Pinnacles National Park:
A Transported Volcanic Landscape

The pinnacles are the eroded remnants of hardened volcanic breccia—slurries of mud and rock from explosive eruptions. This landscape was later plucked from the edge of the North American Plate and transported over 200 miles northwestward along the San Andreas Fault.

to subduct beneath the western edge of North America. This resulted in a line of volcanoes stretching all the way from what is now Alaska to Central America. Beginning about 30 million years ago, so much of this plate was consumed by subduction that the Pacific and North American plates were in contact, forming the San Andreas transform plate boundary in western California. Over time, the San Andreas boundary has grown longer as the Farallon Plate split into two separate plates—the Juan de Fuca Plate on the north, and the Cocos Plate on the south. Remnants of the ancient volcanic mountain chain remain. In central and southern California, for example, the volcanoes have largely eroded away and massive areas of granite from the cooled magma chambers form portions of the Sierra Nevada Mountains, including Yosemite National Park.

The cumulative movement within the broad San Andreas transform plate boundary has had dramatic effects on a landscape that ini-tially developed as part of an ocean/continent subduction zone. For example, rocks found today in Pt. Reyes National Seashore north of San Francisco were originally part of the line of granite rocks formed beneath ancient subduction zone volcanoes. The plate motion has plucked the rocks from their original position and moved them more than 300 miles north-northwestward to their current position at Pt. Reyes. Other rocks in the San Francisco Bay Area were originally part of an "accretionary wedge," similar to those found today in the coastal ranges of the Cascadia Subduction Zone in northern California, Oregon, and Washington.

Modern and Ancient Beasts

Earthquake and volcanic activity occurs at plate boundaries and hotspots for a variety of reasons. At convergent boundaries, where one plate descends (subducts) beneath the other,

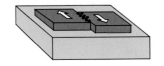

earthquakes are observed over a broad depth range, from near the surface to more than 300 miles below. But earthquakes are generally limited to depths of less than 20 miles at divergent and transform boundaries and hotspots, because cold, rigid plate material (lithosphere) remains at its normal depth. Volcanism occurs at divergent boundaries and hotspots because hot mantle material (asthenosphere) that was under enormous pressure at great depth rises, causing it to decompress and melt. At convergent boundaries crust and mantle material on the subducting plate is subjected to high temperatures and pressures, resulting in the release of fluids and consequent volcanic activity on the surface. But where one plate simply slides past another at a transform boundary, there is little vertical movement of the lithosphere or asthenosphere, and therefore little or no volcanism.

Modern Earthquakes

Two prominent earthquakes along the northern portion of the San Andreas Fault highlight the "Beast" of the West Coast that has wreaked havoc, yet over time has built California's magnificent landscapes and harbors.

The magnitude 7.8 San Francisco Earthquake struck the morning of April 18, 1906. It caused extensive damage to the city, including fires that lasted for several days, and killed an estimated 3,000 people. The earthquake ruptured a large portion of the San Andreas Fault, including land that is now Point Reyes National Seashore and Golden Gate National Recreation Area. The Earthquake Trail at Point Reyes weaves back and forth across the fault line. Exhibits along the trail include the reconstruction of a fence that was offset 16 feet during the 1906 earthquake. Doing some quick math, one can appreciate how dramatically plate-tectonic forces can affect the landscape, even in our lifetimes. The average movement of the Pacific Plate past the North American Plate in California is about two inches per year. If a segment of the San Andreas Fault is "locked" for a cen-

tury, then a large earthquake might result in 200 inches (2 inches/year x 100 years) of movement along the fault in less than a minute. The 16 feet (about 200 inches) of offset along the fence line thus carries a powerful message. Every century or so a large earthquake is necessary to release stress accumulated along large segments of the San Andreas Fault that lock rather than slip smoothly. Knowledge like this helps us better design and site infrastructure, and develop disaster preparedness plans so that our families and communities are less at risk when earthquakes do strike.

On October 17, 1989, just before the World Series baseball game between the Oakland Athletics and San Francisco Giants, a magnitude 6.9 earthquake rattled the Bay Area. Although lessons learned from the 1906 and subsequent quakes helped to lessen the effects, 63 people died, over 3,000 were injured, and property damage exceeded a billion dollars. The earthquake ruptured a segment of the San Andreas Fault that had not moved appreciably since the 1906 quake. One might think that this earthquake was the "one-per-century" event that relieved stress accumulated since 1906. But, alas, the nature of a magnitude scale makes it not so. It takes about 30 earthquakes of magnitude 7 (approximate 1989 quake) to equal the energy released by a magnitude 8 earthquake (approximate 1906 earthquake). The message is that, while the 1989 earthquake was a very significant event, it released only a small part of the stress built up since 1906. The region is still in the window of opportunity for another large earthquake to occur.

Ancient Volcanism

Although the San Andreas Transform Plate Boundary is not currently the site of volcanic activity, there was plenty of that going on in the (geologically) not-too-distant past. From about 200 million to 30 million years ago, Farallon Plate subduction formed a continuous chain of volcanoes from Alaska to Central America.

State Quarters show Iconic Images of National Parks that Represent Modern and Ancient Volcanic Landscapes of the Pacific Northwest

Mt. Rainier National Park, Washington

Mt. Rainier

Robert J. Lillie

Mt. Rainier in Mt. Rainier National Park is an active composite volcano rising more than 14,000 feet above sea level.

Crater Lake National Park, Oregon

Wizard Island

Crater Lake

Robert J. Lillie

Crater Lake in Crater Lake National Park partially fills the large depression formed when a composite volcano erupted and collapsed in on itself 7,700 years ago.

Yosemite National Park, California

Half Dome

Ancient Granite Magma Chambers

Yosemite Valley

Robert J. Lillie

Half Dome in Yosemite National Park is made of granite that solidified from magma tens of millions of years ago, when the subduction zone extended along the entire west coast. The overlying volcanoes have since eroded away, exposing a vast expanse of the ancient magma chambers in the Sierra Nevada.

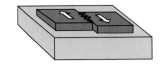
Although most of the volcanoes have eroded from the surface, massive granite outcroppings in the Sierra Nevada are the hardened remnants of ancient magma chambers that fed the volcanoes.

State quarters can be used to demonstrate the current and ancient volcanic landscapes of the Pacific Northwest and California. From 1999 to 2008, the United States Mint issued 25-cent coins representing each of the 50 states. These quarters featured iconic images that have special meaning to each state. Not surprisingly, many of the engravings depict landscapes of national parks. The three West Coast states are especially meaningful because they have geologic features that, when the quarters are stacked, represent the surface and subsurface of the chain of volcanoes that has been developing inland from the coast for the past 200 million years. This can be demonstrated by stacking the three quarters with California on the bottom, overlain by Oregon and then Washington.

Looking first at Washington, one sees Mt. Rainier, a composite volcano rising to over 14,000 feet above sea level in Mt. Rainier National Park. Imagine if Mt. Rainier were to suddenly and violently erupt—so much that its magma chamber empties and the volcanic peak collapses in on itself. Remove the Washington quarter and you can see what the landscape might look like, in the form of Crater Lake in Crater Lake National Park in Oregon. An ancient volcano, Mt. Mazama, erupted and collapsed 7,700 years ago, forming the large cavity (caldera) that now holds Crater Lake. Now suppose the subduction processes that form the volcanoes were to stop, shutting off the magma supply that feeds the volcanoes. With continued uplift and erosion, most of the volcanic material would erode. Remove the Oregon quarter to reveal cooled magma chamber rocks below. This is what has happened in the Sierra Nevada of central and southern California, as represented by the granite-type rocks of Half Dome within Yosemite National Park shown on the California state quarter.

Additional Reading

Alt, David D. and Donald W. Hyndman, 2000, "Roadside Geology of Northern and Central California," Missoula, MT: Mountain Press Publishing Company, 384 pp.

Anderson, D. L., 1971, "The San Andreas Fault," Scientific American, v. 225, no. 5, p. 52-68.

Atwater, Tanya, 1970, "Implications of plate tectonics for the Cenozoic tectonic evolution of western North America," Geological Society of America Bulletin, v. 81, p. 3513-3536.

Collier, Michael, 1999, "A Land in Motion: California's San Andreas Fault," Oakland, CA: University of California Press, 118 pp.

Galloway, A. J., 1977, "Geology of the Point Reyes Peninsula, Marin County, California," California Division of Mines and Geology, Bulletin 202, 72 pp.

Glazner, Allen F., and Greg M. Stock, 2010, "Geology Underfoot in Yosemite National Park," Missoula, MT: Mountain Press Publishing Company, 300 pp.

Hill, M. L., 1981, San Andreas Fault: "History of Concepts," Geological Society of America Bulletin, v. 92, p. 112-131.

Huber, N. King, 1987, "The Geologic Story of Yosemite National Park," Washington, DC: Government Printing Office, U. S. Geological Survey Bulletin 1595, 64 pp.

Iacopi, R., 1996, "Earthquake Country," Fisher Books, 146 pp.

Johnson, E. R., and R. P. Cordone, 1994, "Pinnacles Guide: Pinnacles National Monument, San Benito County, California," Stanwood, Washington: Tillicum Press, 64 pp.

Konigsmark, T., 1998, "Geologic Trips: San Francisco and the Bay Area," Guallala, CA: GeoPress, 174 pp.

McPhee, John, 1993, "Assembling California," New York: Farrar, Straus, and Giroux, 304 pp.

Moore, J. G., 2000, "Exploring the Highest Sierra," Stanford, California: Stanford University Press, 427 pp.

Norris, R. M., and R. W. Webb, 1990, "Geology of California," New York: John Wiley and Sons, Inc., 541 pp.

Schaffer, J. P., 1999, "Yosemite National Park: A Natural History Guide to Yosemite and Its Trails," 4th Edition, Berkeley, California: Wilderness Press, 288 pp.

Schiffman, P., and D. L. Wagner (editors), 1992, "Field Guide to the Geology and Metamorphism of the Franciscan Complex and Western Metamorphic Belt of Northern California," Calif. Div. Mines and Geol., Spec. Pub. 114, 78 pp.

Simpson, R. (editor), 1994, "The Loma Prieta, California Earthquake of October 17, 1989: Tectonic processes and models," U.S. Geol. Surv., Prof. Pap. 1550-F, 131 pp.

Thomas, G., and M. M. Witts, 1971, "The San Francisco Earthquake," New York: Dell Publishing Co., 301 pp.

Trent, D. D., 1984, "Geology of the Joshua Tree National Monument, Riverside and San Bernadino Counties," California Geology, v. 37, p. 75-86.

Trimble, Stephen, 1980, "Point Reyes: The Enchanted Shore, Point Reyes, CA: Coastal Parks Association, 30 pp.

Wallace, R., (editor), 1990, "The San Andreas Fault System, California," U. S. Geological Survey, Professional Paper 1515, 283 pp.

Wells, R. E., C. S. Weaver, and R. J. Blakely, 1998, "Fore-arc migration in Cascadia and its neotectonic significance," Geology, v. 26, p. 759-762.

Wilson, J. T., 1965, "A new class of faults and their bearing on continental drift," Nature, v. 207, p. 343-347.

Yanev, P. I., 1991, "Peace of Mind in Earthquake Country: How to Save Your Home and Life," San Francisco: Chronicle Books, 218 pp.

Yeats, R. S., 2001, "Living with Earthquakes in California: A Survivor's Guide," Corvallis, OR: Oregon State University Press, 406 pp.

4

Hotspot: Columbia Plateau – Yellowstone

The Columbia Plateau has been the site of enormous volcanic eruptions, unsurpassed anywhere on Earth during the past 17 million years. Lava from large fissures in northeastern Oregon and southeastern Washington was so fluid that it flowed considerable distances, forming the numerous layers of basalt familiar to visitors to the Columbia Gorge. Some of the lava traveled more than 300 miles all the way to the Pacific Ocean, where it forms Cape Disappointment, Cape Foulweather, Yaquina

Head, and other headlands notable in the history of Oregon and Washington. This vast volcanism resulted from the rise of hot material from deep within Earth's mantle. Over the past 17 million years the North American continent has continued to drift west-southwest over this hotspot. The spectacular hot springs, geysers, and other hydrothermal features of Yellowstone National Park are the current manifestation of the hotspot activity.

**The Beast of Volcanic Activity has Built the Beauty of the
Columbia Plateau – Yellowstone Hotspot Track**

Columbia Plateau, Oregon

Robert J. Lillie

Yellowstone Plateau, Wyoming

Robert J. Lillie

Columns of basalt (left) represent vast outpourings of fluid lava that covered large portions of Oregon, Washington, and Idaho as the hotspot surfaced 17 million years ago. Geysers, hot springs and other geothermal features in Yellowstone National Park (right) are reminders that the supervolcano that lies directly above the hotspot is still very much alive.

Hotspot Development

Along with black lights and hookahs, lava lamps are iconic fixtures of the psychedelic 1960's. Observing what happens when we turn on a lava lamp can help us understand the evolution of the Columbia Plateau–Yellowstone Hotspot track.

It's not clear exactly why deep mantle material heats up. But when it does it expands like the hot wax in a lava lamp. The wax rises as it becomes less dense than the surrounding oil. At times the rising wax develops a mushroom shape, with a large head and narrow stem. Similarly, deep within the Earth heated mantle becomes less dense than the surrounding material. Although it is still solid, the heated mantle can rise slowly toward the surface. And like the wax in a lava lamp, it can develop a mushroom shape.

Even though Earth's mantle is extremely hot, high pressure deep within the Earth keeps it solid. But at a hotspot, the mantle can rise so fast that it finds itself under much lower pressure while it's still hot. The rising hot mantle eventually begins to melt—much as superheated water expands and flashes to steam when you foolishly take the lid off a pressure cooker!

The magma melting off the mantle at a hotspot initially has a low-silica, basalt composition. When the Yellowstone Hotspot initially reached the surface 17 million years ago, it was shaped like a mushroom, with a large head and narrow stem. The massive outpourings of basalt lava covered the Columbia Plateau surrounding the Washington/Oregon/Idaho juncture, and the Steens Basalt region in southeastern Oregon. The low-silica basalt was so fluid that at times it flowed more than 300 miles through the Columbia Gorge region all the way to the Willamette Valley and Pacific Ocean.

After the initial eruption of the mushroom head, only the narrow stem remained. The thick crust on the overriding plate became more of a factor, because the lesser amounts of magma had to melt their way to the surface. Continental crust is thick and rich in silica. Much of the resulting magma thus had a high-silca, granite composition. This thick, pasty magma can produce explosive eruptions of "super" volcanoes that spew forth lots of pumice and rhyolite lava (the volcanic equivalent of granite). The geothermal features of Yellowstone National Park—and evidence of the latest supervolcano eruption—currently lie directly above the hotspot.

Columbia Plateau

It may seem incredible that lava might flow all the way from near the juncture of Oregon, Washington, and Idaho, through the Columbia Gorge region and all the way to the Willamette Valley and Pacific Ocean. How can the lava stay fluid long enough to travel distances of 300 to 400 miles?

First, consider how fast low-silica lavas can travel. In Hawaii, such lavas have been observed to travel about six miles per hour at the front of the flows where they were moving through thick forests. But once flows cover an area, their surfaces can form channels that carry the lava at speeds of up to 40 miles per hour. And once lava starts to cool and crust over on its surface, it can form a network of lava tubes that distributes lava many miles from the source vents—like blood coursing through your veins! The lava is insulated by the crust of hardened lava, allowing it to remain fluid for a few days.

It's thus conceivable that an individual eruptive event on the Columbia Plateau could spew lava that reached the Coast in just three or four days. (Think about riding your bicycle nonstop at an average speed of six miles per hour. In just three days you would travel more than 400 miles).

The basalt lavas of the Columbia Plateau cover a 63,000 square mile area of Oregon, Washington, and Idaho. Over a span of less than a million years, more than 300 individual flows covered the landscape. In places, these flows add up to a thickness of more than two miles.

The Pacific Northwest shows the Effects of a Tectonic Plate Riding Over a Deep-Mantle Hotspot

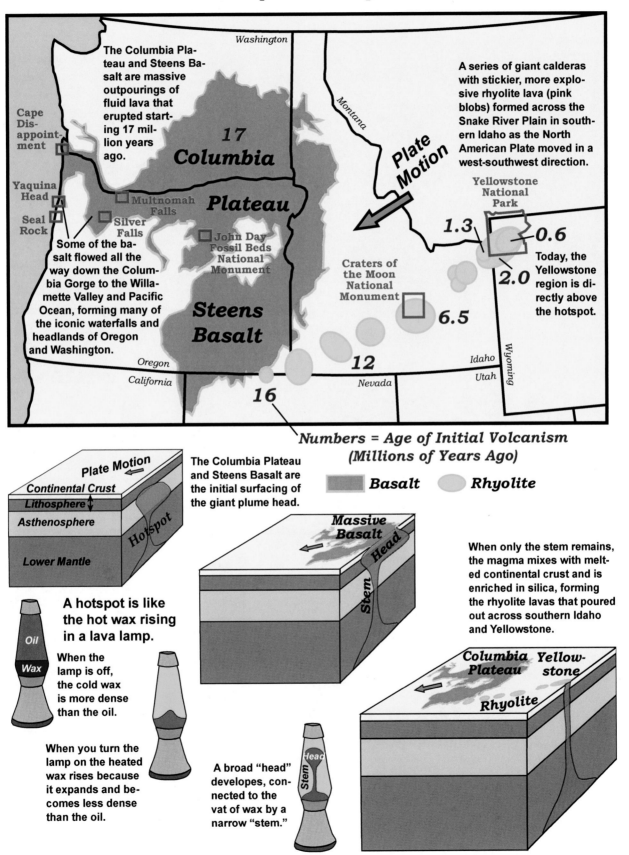

65

Basalt from the Columbia Plateau Forms Many of the Iconic Headlands and Waterfalls of the Pacific Northwest

Cape Disappointment State Park, Washington

Cape Disappointment

Columbia Basalt

Robert J. Lillie

Columbia Basalt

Multnomah Falls

Columbia River Gorge National Scenic Area, Oregon

Robert J. Lillie

Yaquina Head Outstanding Natural Area, Oregon

Yaquina Head

Columbia Basalt

Robert J. Lillie

Seal Rock State Recreation Site, Oregon

Seal Rock

Columbia Basalt

Robert J. Lillie

Silver Falls State Park, Oregon

Columbia Basalt

South Falls

Robert J. Lillie

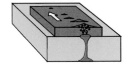

Resistent Lava Flows Form the Caprock of Many Buttes and Mesas in the Columbia Plateau Region

Robert J. Lillie

Sheep Rock in John Day Fossil Beds National Monument, Oregon, is capped by resistent Columbia Plateau Basalt. Softer layers of lake, river, and volcanic ash deposits contain an extraordinary array of horse and other fossils.

Robert J. Lillie

One of the many outstanding murals at the Thomas Condon Visitor Center depicts life on the Columbia Plateau landscape. Note the columnar basalt columns in the foreground.

Robert J. Lillie

Columbia Plateau Basalt often forms columns that are ideally six-sided (hexagons).

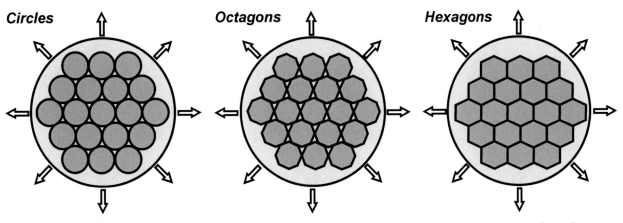

Circles   *Octagons*   *Hexagons*

Developmant of hexagonal columns. As a lava flow cools it shrinks, putting the surface of the flow under tensional forces that form cracks, known as "joints." If the cooling is uniform, circular columns try to form, but there are spaces between the cracks. Octagonal (eight-sided) columns also have spaces. The most efficient form develops hexagonal (six-sided) columns, with no spaces between the cracks.

The total volume of basalt (36,000 cubic miles) is enough to bury the entire lower 48 states under more than 30 feet of lava!

Cooled basalt lava flows are commonly much harder than surrounding sedimentary layers. Consequently, they are responsible for many of the iconic waterfalls and headlands of northwest Oregon and southwest Washington. These features include many oft-visited sites such as Multnomah Falls in the Columbia Gorge, Silver Falls on the eastern edge of the Willamette Valley, and Seal Rock, Yaquina Head, and Cape Disappointment along the Coast.

Yellowstone

The Yellowstone Hotspot track can be thought of as the on land equivalent of the Hawaiian Islands. In Hawaii, the magma from the hotspot stem only has to penetrate the thin crust of the overriding Pacific Plate. Like the rising magma, oceanic crust has a basalt composition. The Hawaiian Islands are thus giant shield volcanoes made of thousands of layers of basalt. And because the Pacific Plate moves northwestward over the hotspot, the islands are progressively older in that direction. The "Big Island" known as Hawaii lies directly over the hotspot and has active volcanoes, including Mauna Loa and Kilauea.

Southern Idaho also has a line of "islands" that formed as the North American Plate moved in a west-southwestward direction over the Yellowstone Hotspot. Starting near the Oregon/Nevada/Idaho juncture 16 million years ago, a line of rhyolite magma centers—supervolcanoes—formed across what is now the Snake River Plain of southern Idaho. Yellowstone National Park today lies directly over the hotspot.

As a plate rides over a hotspot, it rises upward because the underlying mantle beneath is so hot that it expands like a hot-air balloon. The Yellowstone region is thus a high plateau, lying a mile and a half (~8,000 feet) above sea level. This elevation greatly affects climate and ecology. During the last ice age it was covered by a small ice cap, and there are glacial lake

Yellowstone Plateau Development

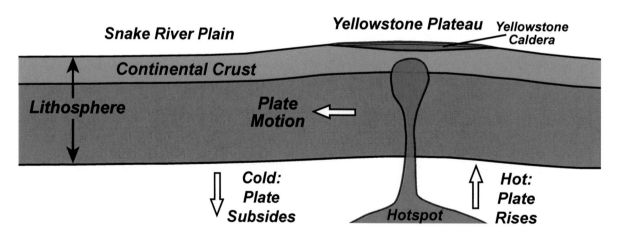

As the Yellowstone Hotspot rises, it expands like a hot-air balloon, causing the Yellowstone Plateau to elevate to great height. Eruptions of silica-rich lava formed the Yellowstone Caldera on the crest of the plateau. The region is so high that a small ice cap—similar to the one that currently covers the island of Iceland—formed during ice ages. The combination of silica-rich soil, high elevations, glacial deposits, and hydrothermal features creates a unique ecosystem on the Yellowstone Plateau. As the plate moves away from the hotspot it cools and contracts, forming lower elevations in the Snake River Plain of southern Idaho.

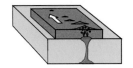

Yellowstone: A Truly Super Volcano!

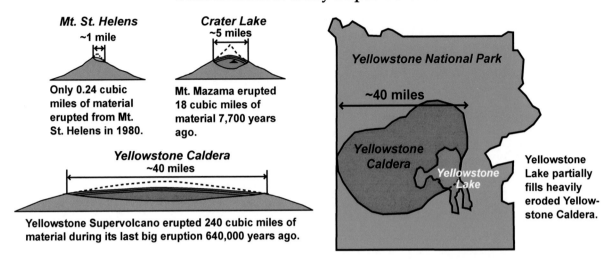

Mt. St. Helens
~1 mile

Only 0.24 cubic miles of material erupted from Mt. St. Helens in 1980.

Crater Lake
~5 miles

Mt. Mazama erupted 18 cubic miles of material 7,700 years ago.

Yellowstone Caldera
~40 miles

Yellowstone Supervolcano erupted 240 cubic miles of material during its last big eruption 640,000 years ago.

Yellowstone National Park

~40 miles

Yellowstone Caldera

Yellowstone Lake

Yellowstone Lake partially fills heavily eroded Yellowstone Caldera.

The giant crater ("caldera") formed by the eruption and collapse of Yellowstone Supervolcano 640,000 years ago dwarfs the crater on top of Mt. St. Helens and the caldera of Mt. Mazama.

Grand Canyon of the Yellowstone

Yellowstone National Park, Wyoming

Robert J. Lillie

The Yellowstone River cuts through rhyolite lava flows that partially fill the Yellowstone Caldera.

69

Development of Yellowstone Hydrothermal Features

2. Magma that encounters silica-rich continental crust on its journey upward forms a rhyolite magma chamber only five to ten miles beneath Yellowstone National Park.

3. Water from rainfall and snowmelt seeps into the ground.

5. Hot water reaches the surface as geysers, hot springs, mudpots, and fumaroles.

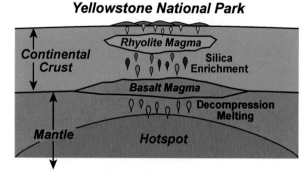

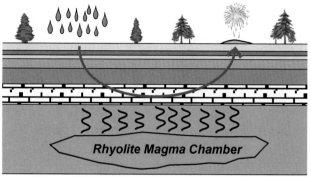

1. Hot material rises from deep within Earth's mantle and melts, forming basalt magma at the base of the crust.

4. The water circulates one to two miles underground, where it encounters hot rock above the rhyolite magma chamber.

Yellowstone Hydrothermal Features:
Fumaroles, Mudpots, Paint Pots and Mud Volcanoes

Dragon's Cauldron

Mud Volcano

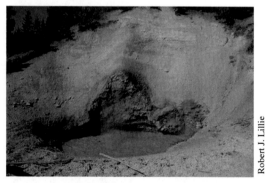

Artists Paint Pots

Mudpots, paint pots and mud volcanoes develop where a lot of fine rock material is incorporated into the hot water. Fumaroles are eruptions of steam on the surface. Dazzling colors occur where steam and hot water alter minerals.

Yellowstone Hydrothermal Features:
Hot Springs and Geysers

Fishing Cone

Robert J. Lillie

Heart Spring

Robert J. Lillie

Blue Star Spring

Robert J. Lillie

Hot Springs develop where the geothermal waters flow freely to the surface.

Mammoth Hot Springs

Robert J. Lillie

Mammoth Hot Springs

Robert J. Lillie

Mammoth Hot Springs form in the northeast part of the park where the hot water flows through limestone. Dissolved calcium carbonate is deposited on the surface as beautiful terraces.

Old Faithful Geyser

Robert J. Lillie

Sawmill Geyser

Robert J. Lillie

Geysers form where dissolved silica from rhyolite volcanic rocks constricts flow. Under pressure, the superheated water flashes from liquid to gas, propelling the column of water upward through the constriction.

71

Life Above a Hotspot

Robert J. Lillie

Robert J. Lillie

Robert J. Lillie

Yellowstone's high elevations and silica-rich volcanic soils favor lodgepole pine forests. Glacial lake deposits in Hayden Valley form wetland meadows ideal for bison and other wildlife. Bison, elk and grizzly bears thrive in the forests and meadows.

Robert J. Lillie

Robert J. Lillie

Robert J. Lillie

Bison hang around open areas near geothermal features. The rich yellow and orange colors around many hot springs are microscopic thermophiles—special forms of life that survive at very high temperatures.

Robert J. Lillie

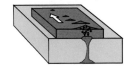

deposits. Lodgepole pines dominate the forests because, unlike most trees, they thrive on the stressful conditions of high altitude and acidic soils formed from silica-rich pumice and rhyolite.

When the plate moves off the hotspot, the mantle beneath cools and contracts, causing the region to sink down. This is evident in Hawaii, where islands get shorter in a northwestward direction, eventually sinking beneath the sea as atolls and seamounts. Similarly, the Snake River Plain of southern Idaho is a depression in the landscape where the North American Plate has moved off the Yellowstone Hotspot. Younger basalt lava flows (some from Basin and Range continental rifting) cover much of the earlier rhyolite supervolcanic centers.

Additional Reading

Alt, David D. and Donald W. Hyndman, 1984, "Roadside Geology of Washington," Missoula, MT: Mountain Press Publishing Company, 288 pp.

Alt, David D. and Donald W. Hyndman, 1989, "Roadside Geology of Idaho," Missoula, MT: Mountain Press Publishing Company, 403 pp.

Anderson, D. L., T. Tanimoto, and Y. Zhang, 1992, "Plate tectonics and hotspots: The third dimension," Science, v. 256, p. 1645-1651.

Brock, T. D., 1994, "Life at High Temperatures," Yellowstone Association for Natural Science, History and Education, Inc., 31 pp.

Bryan, T. S., 1990, "Geysers: What they are and how they work," Niwot, Colorado: Roberts Rinehart, Inc., 24 pp.

Christopherson, E., 1962, "The Night the Mountain Fell: The Story of the Montana-Yellowstone Earthquake," West Yellowstone, Montana: Yellowstone Publications, 88 pp.

Decker, R., and B. Decker, 2001, "Volcanoes in America's National Parks," New York: W. W. Norton and Comp., 256 pp.

Foley, Duncan, 2006, "Yellowstone's Geysers: The Story Behind the Scenery," Las Vegas, NV: KC Publications, 48 pp.

Good, J. M., and K. L. Pierce, 1996, "Interpreting the Landscape: Recent and Ongoing Geology of Grand Teton and Yellowstone National Parks," Moose, Wyoming: Grand Teton Natural History Association, 58 pp.

Fritz, William J. and Robert C. Thomas, 2011, "Roadside Geology of Yellowstone Country," Missoula, MT: Mountain Press Publishing Company, 2nd Edition, 328 pp.

Keefer, W. R., 1971, "The geologic story of Yellowstone National Park," U. S. Geol. Surv., Bull. 1347, 92 pp.

Miller, Marli B., 2014, "Roadside Geology of Oregon," 2nd Ed, Missoula, Mt: Mountain Press Publishing Company, 380 pp.

National Park Service, 1991, "Craters of the Moon: Official National Park Handbook," Washington, DC: U.S. Government Printing Office, Handbook Number 139, 64 pp.

Orr, Elizabeth L. and William N. Orr, 2000, "Oregon Geology," 6th Edition, Corvallis, OR: Oregon State University Press, 304 pp.

Smith, R. B., and R. L. Christiansen, 1980, "Yellowstone Park as a window on the Earth's interior," Scientific American, v. 242, p. 104-117.

Smith, Robert B., and Lee J. Siegel, 2002, "Windows into the Earth: A Geologic Story of Yellowstone and Grand Teton National Parks," New York: Oxford University Press, 242 pp.

Spearing, Darwin, and David Lageson, 1988, "Roadside Geology of Wyoming," Missoula, MT: Mountain Press Publishing Company, 288 pp.

Thayer, T. P., 1994, "Geologic Setting of the John Day Country," U. S. Geological Survey, 23 pp.

Williams, Ira A., 1991, "Geologic History of the Columbia River Gorge," Portland, OR: Oregon Historical Society, 137 pp.

Glossary

absolute plate motion – The speed and direction a tectonic plate moves over the fixed, deep portions of the Earth (for example, over a **hotspot**).

accretionary wedge – A mountain range formed as sedimentary layers and hard crust are scraped off the top of a subducting plate.

accreted terrane – A block of crust (commonly oceanic islands or continental fragments) that was added onto the edge of a continent due to plate-tectonic movements.

andesite – A fine-grained, light-to-dark colored **igneous rock** with about 60% **silica**.

ash – (See **volcanic ash**).

asthenosphere – Relatively soft portion of Earth's upper **mantle**. The rigid plates of **lithosphere** drift about over the flowing asthenosphere.

basalt – A fine-grained, dark-colored **igneous rock** with about 50% **silica**.

basaltic andesite – A fine-grained **igneous rock** with about 55% **silica**. (More silica than basalt, but less than andesite).

basin - A depression that accumulates sedimentary deposits.

Basin and Range Province – A region of long, north-south trending mountain ranges and intervening valleys in the western United States, formed by ongoing continental rifting.

batholith – An extensive, igneous **intrusive rock** body, commonly composed of **granite** or **granodiorite**. Typically forms where many **magma chambers** cool.

breccia – 1. A **sedimentary rock** made up of coarse, angular fragments. (A similar rock with coarse, rounded fragments is a **conglomerate**). 2. A **volcanic rock** with coarse, angular rock fragments encased in finer volcanic particles.

brittle – Failure of solid material by discrete cracking and breaking (like peanut brittle or cold plastic).

bomb – See **volcanic bomb**.

buoyancy – The tendency of material to rise because it is less dense that the surrounding material.

calcite – A mineral composed of the compound calcium carbonate ($CaCO_3$). The primary mineral comprising the sedimentary rock **limestone**, and its metamorphic equivalent, **marble**.

caldera – A relatively large, steep-sided crater, at least ½ mile (1 kilometer) in diameter, formed by the eruption and collapse of a volcano.

Cascadia Subduction Zone – A region of parallel coastal and volcanic mountain ranges in the **Pacific Northwest**, formed by plate convergence where the offshore **Juan de Fuca Plate** dives (**subducts**) beneath the edge of the **North American Plate**.

cinders – Pea-to-gravel sized particles formed as molten material ejected during a volcanic eruption

← Crater Lake National Park, Oregon. Glaciers formed U-shaped valleys on a once-lofty volcano, Mt. Mazama. (Photo by Robert J. Lillie).

cools and solidifies in the air. Commonly has the chemical composition of **basalt** or **basaltic andesite**. (See **scoria**).

cinder cone – A relatively small, < 1 mile (1½ kilometer) diameter volcano, formed as **pyroclastic** material (volcanic bombs, cinders, ash) blasts up into the air and rains down as solid particles that pile up into a cone with a slope of about 30°. Its materials commonly have the chemical composition of **basalt** or **basaltic andesite**.

clastic sedimentary rock – A rock formed from the eroded fragments of other rocks.

column – An upright structure bounded by cracks (**joints**) formed as a **lava flow** cools.

columnar jointing – Cracks formed due to the shrinking of a lava flow as it cools. If cooling is uniform, the cracks commonly result in vertical, six-sided (hexagonal) columns.

composite volcano – A large (up to 12 mile; 20 kilometer diameter) mountain formed through numerous eruptions of a variety of volcanic materials (lava flows, ash, pumice, cinders, mudflows) that pile up into a relatively steep volcano (also known as a **stratovolcano**).

Columbia Basalt – Lava rock that erupted from fissures in the **Columbia Plateau**, mostly from 17 to 15 million years ago. The numerous lava flow layers are collectively known as the "Columbia River Basalt." Along with the **Steens Basalt**, these rocks are thought to represent the initial surfacing of the **Yellowstone Hotspot**.

Columbia Plateau – A broad region of southeast Washington, northeast Oregon, and western Idaho that was covered with massive outpourings of **basalt** lava, starting 17 million years ago and gradually diminishing over the next 10 million years.

conglomerate – A sedimentary rock formed from the compaction and cementation of sediment containing a lot of pebbles or cobbles. (If those particles are angular, the rock is called **breccia**).

continental collision zone – A region of high topography formed as an ocean basin closes and thick crustal blocks crash into one another. Occurs at a **convergent plate boundary**, but differs from a **subduction zone** in that both plates have thick continental crust. A modern example is the Himalayan Mountains and Tibetan Plateau, where the Indian subcontinent is crashing into Asia.

continental craton – (See **craton**).

continental crust – Outer layer of the Earth that forms the continents. It is thicker and higher in **silica** than **oceanic crust**.

continental drift – The theory that continents are not stationary, but rather move about relative to one another.

continental rift – A region of long mountain ranges and valleys formed as a plate capped by thick continental crust is pulled apart by plate divergence.

convergent plate boundary – Region where two slabs of Earth's outer shell (**lithosphere**) move toward one another, destroying lithosphere. **Subduction zones** and **continental collision zones** are common manifestations of convergent plate boundaries.

core – The central region of the Earth composed mostly of iron and nickel.

craton – A relatively flat-lying, older region of a continent that has not experienced widespread tectonic activity for hundreds of millions of years.

crust – The outermost part of the Earth composed mostly of light **silicate** minerals.

dacite – A fine-grained **igneous rock** with about 65% silica. (More silica than **andesite**, but less than **rhyolite**).

decompression melting – A change from solid to liquid that occurs as the pressure on hot, pressurized material drops. Occurs as hot mantle material rises quickly.

dehydration melting – A change from solid to liquid that occurs when a material is heated and looses water (dehydrates). Occurs where a plate subducts and loses water as it heats up; the water rises and melts rock in its path.

deposition – The accumulation of sedimentary material transported by water, wind or ice.

dike – A sheet-like, igneous **intrusive rock** body formed where **magma** cuts across rock layers.

diorite - A coarse-grained, light-to-dark colored **igne-**

ous rock with about 60% **silica**. (Intrusive equivalent of **andesite**).

divergent plate boundary – A region where two slabs of Earth's outer shell (**lithosphere**) pull apart from one another, creating new lithosphere. **Mid-ocean ridges** and **continental rifts** are common manifestations of divergent plate boundaries.

ductile - Failure of solid material by flowing (like Silly Putty or hot plastic).

earthquake – A sudden movement within the Earth that releases vibrations (**seismic waves**).

earthquake intensity – A number describing the severity of actual effects due to earthquake shaking at a particular place on Earth's surface.

earthquake magnitude – A measure of the amount of seismic energy released by an earthquake. The scale is logarithmic, such that an increase in one number on the magnitude scale corresponds to a ten-fold increase in the amplitude of **seismic waves**, and about a 30-fold increase in the amount of energy released.

earthquake recurrence interval – The average time between earthquakes of a given **magnitude** along a particular **fault** or fault segment.

ecology – The relationship between organisms and their environment.

ecosystem – A community of organisms and their physical environment, considered as a unit.

erosion – The breakdown of rock into smaller particles by the mechanical actions of water, wind or ice, or by biological or chemical activity.

escarpment – (See **fault escarpment**).

extrusive rock - An **igneous rock** that solidified from magma that erupted on Earth's surface, forming fine-grained mineral crystals (**volcanic rock**).

fault – A break in the Earth along which the blocks on either side have moved parallel to the break.

fault-block mountains – Long, narrow ranges that moved upward relative to adjacent valleys along faults.

fault escarpment – A change in elevation representing the part of a **fault** extending above the surface.

feldspar – One of a group of magnesium/aluminum-silicate minerals that are common in rocks in Earth's **crust**.

forearc basin – A depression between the uplifted (**accretionary wedge**) and volcanic (**volcanic arc**) mountain ranges formed above a **subduction zone**.

formation – A mass of rock with recognizable characteristics that make it distinguishable from surrounding masses of rock. Usually applied to a layer or sequence of layers of **sedimentary rock**.

fossil – The remnants of an ancient plant or animal. The original material might be replaced by rock, or an impression left in the rock, as the plant or animal decays.

Franciscan Group – Remnants of an ancient **accretionary wedge**, developed during the subduction of the Farallon Plate and preserved in western California.

fumarole – A hole or crack in a volcanic region through which gases and vapors rise.

GPS – (Global Positioning System) - A group of orbiting satellites used to find the location of a point on Earth's surface.

gabbro – A coarse-grained, dark-colored **igneous rock** with about 50% **silica**. (Intrusive equivalent of **basalt**).

geological hazard – A natural situation on or within the Earth that is capable of causing property damage, injury, or loss of life. Examples include conditions that might produce an **earthquake**, **landslide**, **tsunami**, or **volcanic eruption**.

geology – The study of the Earth.

geomorphology – The study of landforms, like mountains, valleys, and shorelines, and the processes that formed them.

geosynclinal theory – The idea that mountain building, volcanic eruptions, metamorphism, and other **tectonic** activity involves the downwarping of large regions of Earth's crust into giant depressions, or synclines. Suggests that tectonic features are due mainly to vertical movements of Earth's crust. Contrasts with the modern **plate tectonic theory**, which relies more on horizontal motions.

geothermal gradient – The rate of increase of temperature with depth within the Earth.

geyser – A spout that periodically erupts water from the Earth.

gneiss – A **metamorphic rock** formed from a pre-existing rock that underwent extreme increases in temperature and pressure. (Greater amount of metamorphism than **schist**).

Gorda Plate – The southern part of the **Juan de Fuca Plate** off the coast of northern California.

Gorda Ridge – A **divergent plate boundary (mid-ocean ridge)** where the **Gorda Plate** moves away from the **Pacific Plate**.

granite - A coarse-grained, generally light-colored **igneous rock** with about 70% **silica**.

granodiorite – A coarse-grained **igneous rock** with about 65% **silica**. (Less silica than **granite**, but more than **diorite**).

hot spring – A place where hot water escapes from the ground.

hotspot – Plume of hot material that rises from Earth's deep **mantle**. Lines of volcanoes form on the surface where a tectonic plate rides over a hotspot.

hydrothermal alteration – A change in the mineral composition of rock induced by hot water.

igneous rock – A rock that solidified from molten Earth material (**magma**).

inner core – The zone of dense, solid iron and nickel extending from 3,200 miles (5,100 kilometers) depth to the Earth's center at 4,000 miles (6,300 kilometers).

intensity – (See **earthquake intensity**).

intrusive rock – An **igneous rock** solidified from **magma** that cooled within the Earth, generally forming coarse-grained mineral crystals (**plutonic rock**).

Juan de Fuca Plate – A part of Earth's outer shell (**lithosphere**) consisting of a small portion of the Pacific Ocean off the coast of the **Pacific Northwest**.

Juan de Fuca Ridge – A **divergent plate boundary**

(**mid-ocean ridge**) where the **Juan de Fuca Plate** moves away from the **Pacific Plate**.

joint – A crack in a mass of rock, with slight opening but no significant lateral movement of the blocks of rock on either side of the crack. (A crack involving significant lateral movement is a **fault**).

lahar – 1. A mass of solid material and water that moves swiftly downslope, commonly as a result of volcanic processes (**volcanic mudflow**). 2. The deposits of a volcanic mudflow.

landslide – The downward movement of a mass of rock, soil, and other natural or artificial materials that suddenly break free from a mountainside or other steep slope.

landscape – An area of Earth's surface that has a particular type of appearance.

lava – Hot, molten rock (**magma**) that poured out on Earth's surface or beneath the sea.

lava dome – A relatively small volcano, < 1 mile (1½ kilometers) in diameter, formed by the eruption of sticky, silica-rich magma (**rhyolite**, **dacite**) around a central vent.

lava flow – 1. Molten material (**magma**) pouring out on Earth's surface. 2. Hard rock formed from magma that erupted from the summit of a volcano, or from fissures along its sides.

lava tube – A long, commonly smooth cave formed where lava crusts over and the remaining lava flows out.

limestone – A **sedimentary rock** formed from the compaction and cementation of calcium carbonate (the mineral **calcite**) that precipitated out of water.

lithosphere – The rigid outer shell of the Earth, composed of the outermost **mantle** and **crust**. The lithosphere is broken into plates that move over the underlying, softer **asthenosphere**.

magma – Hot, liquid rock that may contain some gas and solid material.

magma chamber – An accumulation of molten rock below Earth's surface.

magnitude – (See **earthquake magnitude**).

mantle – The portion of the Earth between the **crust** and **core**, consisting of **silicates** rich in iron and magnesium.

marble – A **metamorphic rock** formed from **limestone** that underwent extreme increases in temperature and pressure.

metamorphic rock – A rock formed through the re-crystallization of a preexisting rock, while the rock was still solid.

metamorphose – To change from one form to another. (Commonly applied to life cycles of insects, and to changes that occur when rock is subjected to increased temperature and pressure).

microcontinent – A small block of continental crust that broke off a larger continent.

mid-ocean ridge – An undersea mountain range formed from volcanic activity where tectonic plates capped by thin **oceanic crust** move away from one another (**divergent plate boundary**).

mineral – A naturally occurring, inorganic solid with specific chemical composition and crystalline structure.

mountain range – A region of high topography formed by crustal deformation or volcanic processes.

mudflow – (See **volcanic mudflow** or **lahar**).

mudpot – A **hot spring** with water that has incorporated a lot of Earth material on its way to the surface.

North American Plate – A large part of Earth's outer shell (**lithosphere**) that includes most of the North American continent and the western portion of the Atlantic Ocean.

obsidian – An **igneous rock** with glassy texture, formed when **magma** cools so quickly that it does not have time for mineral crystals to develop.

oceanic crust – The outer layer of the Earth underlying the oceans. It is thiner and lower in **silica** than **continental crust**.

ophiolite – The rock sequence comprising **oceanic crust** and the uppermost **mantle**. From bottom to top an ophiolite consists of the igneous rocks **peridotite**, **gabbro**, and **basalt**, and a deep-ocean **sedimentary rock** layer.

outer core – The zone of dense, liquid iron and nickel extending from 1,800 miles (2,900 kilometers) depth to 3,200 miles (5,100 kilometers) between the solid **lower mantle** and solid **inner core**.

Pacific Northwest – A broad region including all of Oregon and Washington, adjacent portions of Idaho, California, and British Columbia, and sometimes parts of Nevada, Montana and Wyoming.

Pacific Plate – A large part of Earth's outer shell (**lithosphere**) that includes most of the Pacific Ocean region and pieces of adjacent continents.

peridotite - A coarse-grained, olive-green colored **igneous rock** with about 40% **silica**. (Constitutes most of Earth's **mantle**).

pillow basalt – (See **pillow lava**).

pillow lava – A rock layer (**lava flow**) with globular structures formed when magma erupted through the sea floor or flowed into the ocean or other body of water. (Commonly called **pillow basalt** because the free-flowing lava usually has **basalt** to **basaltic andesite** composition).

plate tectonics – The theory that Earth's major surface features, including continents, oceans and mountain ranges, result from the horizontal movement of large plates of Earth's outer shell (**lithosphere**).

plutonic rock – An **igneous rock** that solidified from **magma** that cooled within the Earth (**intrusive rock**).

primary magma – Molten material that initially forms from melting of a solid portion of the Earth. Commonly refers to magma of **basalt** composition that melts off the **peridotite** of the **mantle**. The composition of a primary magma can be altered as it rises through the **crust** and incorporates other materials.

pumice – A low-density **igneous rock**, rich in **silica** and pore spaces, formed from frothy **lava**.

pyroclastic – Material flung into the air during a volcanic eruption.

quartz – A **mineral** composed of the elements silicon and oxygen (chemical formula SiO_2).

quartzite – A **metamorphic rock** formed from a **quartz**-rich **sandstone** layer that underwent extreme increases in temperature and pressure.

recurrence interval – (See **earthquake recurrence interval**).

relative plate motion – The speed and direction of one plate compared to another at the boundary of the two plates.

rhyolite – A fine-grained, generally light-colored **igneous rock** with about 70% **silica**. (Extrusive equivalent of **granite**).

Ring of Fire – The region bordering the Pacific Ocean with high levels of earthquakes and volcanic eruptions, primarily due to activity at **subduction zones**.

rift valley – A depression between mountain ranges in a **continental rift**. (See **basin**).

rift valley strata – Sedimentary and volcanic layers deposited in a **rift valley**.

rock – A mixture of **minerals** that are cemented together in some natural way.

San Andreas Fault – A major break in Earth's crust extending from northern Mexico through western California. Lateral movement of the landscape along the fault line is part of the north-northwestward movement of the **Pacific Plate** past the **North American Plate** along a **transform plate boundary**.

sandstone – A **sedimentary rock** formed from the compaction and cementation of eroded rock particles the size of sand. (Commonly rich in the mineral **quartz**).

schist – A **metamorphic rock** formed from a pre-existing rock that experienced relatively large increases in temperature and pressure. (Greater amount of metamorphism than **slate**, but less than **gneiss**).

scoria – A porous **igneous rock**, low in **silica** and rich in iron; more dense than **pumice**. Commonly forms volcanic **cinders** and **bombs**.

seamount – An undersea mountain formed as a volcano erodes and subsides.

sedimentary basin – A depression that holds water and accumulates deposits of rock fragments and other sedimentary materials.

sedimentary rock – A **rock** formed from the burial and cementation of eroded rock fragments, or from material created through biological or chemical activity.

seismic wave – Sound or other vibration that moves through the Earth or along its surface, caused by an **earthquake**, **landslide**, or other natural or artificial source.

seismology – The study of **earthquakes** and the waves they generate.

shale – A **sedimentary rock** formed from the compaction and cementation of fine mud.

shearing – The change in shape of a material as it is stressed. Can be demonstrated by holding a deck of cards in your hands and sliding them laterally—shearing occurs where the cards slip along their faces.

shield volcano – A broad, up to 60-mile (100-kilometer) diameter mountain, with gentle slopes formed through the eruption of numerous fluid (**basalt** or **basaltic andesite**) lava flows.

Sierra Nevada – A north-south trending mountain range in eastern California and its border with Nevada. Consists of granitic and metamorphic rocks that are the remnants of an ancient **volcanic arc**.

sill – A sheet-like, igneous **intrusive rock** body formed where **magma** cools parallel to rock layers.

silica – An ion consisting of the elements silicon and oxygen that combines with other elements to form most of the **minerals** in Earth's **crust** and **mantle**. Pure silica forms the mineral **quartz** (chemical formula SiO_2).

silicate – A compound that includes the elements silicon and oxygen, with one atom of silica for every two oxygen atoms.

siltstone - A **sedimentary rock** formed from the compaction and cementation of eroded rock particles finer than sand but more coarse than clay.

slate – A **metamorphic rock** formed from a **shale** layer that experienced relatively small increases in temperature and pressure. (Lesser amount of metamorphism than **schist**).

spatter cone – A mound of hardened lava, from a few to several feet (meters) across. Typically forms as fluid (**basalt**) lava erupts in blobs that fly a short distance and accumulate around a volcanic vent.

Steens Basalt – **Massive** outpourings of **basalt** lava that erupted in the Steens Mountain area of southeastern Oregon about 17 million years ago. Along with similar **lava flows** in the **Columbia Plateau**, these rocks are thought to represent the initial surfacing of the **Yellowstone Hotspot**.

stratovolcano – (See **composite volcano**).

subduct – To extend beneath and disappear. Occurs at a **convergent plate boundary** where at least one of the plates is capped by **oceanic crust**.

subduction zone – A **convergent plate boundary** where a plate capped by **oceanic crust** slides beneath another plate and extends deeply into Earth's **mantle**.

tectonic setting – The type of plate boundary or hotspot responsible for the formation of the rocks and topography of a region.

tectonics – The study of large features on Earth's surface and the internal processes that led to their formation.

tephra – **Pyroclastic** material of any size ejected from a volcano or fissure during explosive activity.

terrane – An extensive region that is bounded by **faults** and has distinctive geology that differs considerably from the geology of surrounding regions. (See **accreted terrane**).

thermal subsidence – The downward movement of a region caused by the contraction of rock as it cools.

thermophilic bacteria – Organisms capable of surviving at temperatures near or above the boiling point of water.

transform plate boundary – The region where two slabs of Earth's outer shell (**lithosphere**) slide laterally past one another.

tsunami – A series of giant sea waves caused by movement of the sea floor due to an earthquake, landslide, or volcanic flow. (Sometimes mistakenly called a "tidal wave").

vesicular basalt – **Basalt** that has visible air pockets formed where gas expanded in cooling **magma**.

volcanic arc – The chain of volcanoes that forms on the overriding plate at a **subduction zone**.

volcanic ash – Fine-grained **pyroclastic** material, blown from a volcano and carried away and deposited by winds.

volcanic breccia – **Volcanic rock** composed of coarse, angular fragments within a finer-grained mass.

volcanic bomb – A large piece of **pyroclastic** material, commonly football to watermelon size, that developed the shape of an artillery or airplane bomb as it traveled trough the air in a liquid state.

volcanic cinders – (See **cinders**).

volcanic eruption – The sudden release of volcanic materials (commonly a mixture that might include **ash**, **magma**, and other gases, liquids and solids) from a volcano or fissure in the Earth.

volcanic mudflow – 1. A mass of solid material and water that moves swiftly downslope, commonly as a result of volcanic processes (**lahar**). 2. The deposits of a volcanic mudflow.

volcanic rock – An **igneous rock** that solidified from **magma** that erupted on Earth's surface (**extrusive rock**).

volcano – A commonly cone-shaped mountain or hill through which molten Earth materials erupt.

volcanology – The study of molten material erupted from the Earth and the products of such eruptions.

Yellowstone Hotspot – Hot material rising from deep within Earth's Mantle to the region of the **Yellowstone Plateau**.

Yellowstone Plateau – A region of high elevation centered around Yellowstone National Park of northwestern Wyoming and adjacent portions of Idaho and Montana.

About the Author

Robert J. (Bob) Lillie is a free-lance writer, illustrator, and interpretive trainer, specializing in communicating park landscapes and their deeper meanings to the public. Bob was a Professor of Geosciences at Oregon State University from 1984 to 2011, where he taught courses in physical geology, oceanography, tectonics, geophysics, geological writing, and public interpretation. He is author of *Parks and Plates: The Geology of Our National Parks, Monuments, and Seashores* (W. W. Norton and Company, 2005) and is a Certified Interpretive Trainer (CIT) through the National Association for Interpretation (NAI). From 2007 to 2011 Bob was the Manager of Education and Outreach for EarthScope, a nationwide program of the National Science Foundation. From 2012 to 2014 he served with the Cascadia EarthScope Earthquake and Tsunami Education Program to train teachers, interpreters, and emergency management educators on geological hazards and preparedness in the Pacific Northwest.

Bob was born and raised in the Cajun Country of Louisiana. He has a B.S. in geology from the University of Louisiana–Lafayette, an M.S. in geophysics from Oregon State University, and a Ph.D. in geophysics from Cornell University. Bob's research focuses on the geological evolution of mountain ranges formed by the collision of continents, including the Himalayas in India and Pakistan and the Carpathians in Central Europe. He is also author of *Whole Earth Geophysics: An Introductory Textbook for Geologists and Geophysicists* (Prentice Hall, 1999), used in college courses in the United States and other countries.

Since 1994 Bob has collaborated with the National Park Service (NPS) on educating the public in geology. He served as a seasonal interpretive ranger at Crater Lake and Yellowstone national parks and John Day Fossil Beds National Monument, and he and his graduate students have written and illustrated geology-training manuals for NPS sites across the country. Bob was presented the 2005 NPS Geological Resources Division award for "outstanding contributions in engaging the National Parks staff and visitors in geoscience."

Bob has done bicycle tours of the U.S., Ireland, the Alps, Central Europe, and Scandinavia, and his hobbies include photography and Cajun cooking. He lives with his wife Barb on Wells Creek in the Oregon Coast Range west of Philomath.

← Wells Creek near Marys Peak, Oregon. Sandstone layers deposited in the Pacific Ocean have been uplifted as part of the Coast Range. (Photo by Robert J. Lillie).

Beauty from the Beast
Plate Tectonics and the Landscapes of the Pacific Northwest

Robert J. Lillie, PhD, Certified Interpretive Trainer
Emeritus Professor of Geosciences
College of Earth, Ocean, and Atmospheric Sciences
104 CEOAS Admin Bldg, Oregon State University
Corvallis, OR, 97331
Phone: (541)-231-2247
E-mail: lillier@geo.oregonstate.edu
Web: www.robertjlillie.com

Beauty from the Beast introduces readers to the story behind the scenery and geological hazards of the Pacific Northwest and surrounding regions. It includes:

- A complete treatment of plate tectonics and landscape formation in the greater Pacific Northwest, with focus on national and other parklands for each of the three types of plate boundaries and a hotspot (the "Beauty").

- Details of how earthquakes, tsunamis, and volcanic eruptions are caused by the same plate-tectonic forces (the "Beast").

- Color photographs of landscape features in national, state, and local parklands that depict specific geological features and processes.

- Illustrations of plate-tectonic processes responsible for park landscapes and geological hazards, including many with "Fun-with-Food" and other devices to enliven the text and connect to readers via experiences from their everyday lives.

Back Cover: Cape Perpetua Scenic Area reveals the beauty of the central Oregon Coast (photo by Robert J. Lillie). The tsunami evacuation sign, developed in the Pacific Northwest and used worldwide, helps communities live with the beast of the Cascadia Subduction Zone.

Made in the USA
Columbia, SC
25 July 2017